O&B
MATHS BANK
3

CHELTENHAM GRAMMAR SCHOOL
MATHEMATICS DEPARTMENT

O&B
MATHS BANK
3

K. J. Dallison M.A.

J. P. Rigby B.Sc.

Oliver & Boyd

Ilustrated by David T. Gray and Tom Reid

Oliver & Boyd
Robert Stevenson House
1–3 Baxter's Place
Leith Walk
Edinburgh EH1 3BB

A Division of Longman Group Ltd.

ISBN 0 05 003156 2

First published 1979
Sixth impression 1986

Produced by Longman Group (FE) Ltd
Printed in Hong Kong

Contents

Foreword

I first came across the authors' enthusiasm as teachers ten years ago, when they founded our local mathematics association for sixth formers and invited me to become its first president.

All mathematics teachers who use one of the newer types of school syllabus will recognise the usefulness of these books. The introduction of any new syllabus should always be backed up by a wealth of problems, not only for the teacher's benefit, but also to provide side paths along which the pupils can explore and play. This is especially true in the case of the more gifted, and especially in the mixed ability classes of today, in which the more gifted must necessarily be expected to work more on their own.

Learning (that is assimilating and memorising) any syllabus can be a tiresome business, and in mathematics at any level there are always opportunities for indulging in the more challenging and exciting tasks of discovery and creativity as well. These require reflection alone, and some form of guidance. There is no better guidance than well chosen problems that appeal to the intuition and focus the imagination, and through which the student can recreate his or her own mathematics. Such self-discovery leads to a much deeper understanding, and a confidence in the subject, which the student will never forget and upon which he or she can build further.

E. C. ZEEMAN, F.R.S.
University of Warwick

Preface

These books are rather different from the usual mathematics texts in that they contain almost no teaching material. They do, however, contain a wealth of questions, covering the modern and traditional mathematics required by 'O' level and C.S.E. syllabuses in modern mathematics.

Maths Bank can be used to supplement any existing course in modern mathematics but it can also be used as a course book in its own right, leaving the teacher free to instruct in his own way.

The questions are designed to cater for a wide range of ability: each section begins with easier questions; harder and deliberately wordy questions are starred.

The authors wish to express their gratitude to Miss P. M. Southern, Mr E. P. Willin and other past and present members of the mathematics staff at Rugby High School for writing questions and supplying answers; to Mr M. E. Wardle, head of the Department of Mathematics at Coventry College of Education for acting as adviser on difficult points; and to Miss D. M. Linsley, former headmistress of Rugby High School, without whose foresight in allowing the school to change to modern mathematics in 1963 these books would never have been written.

K. J. DALLISON
J. P. RIGBY
Rugby High School

1 Simultaneous Linear Equations

1A

Solve the following, finding values for both x and y in each pair of equations.

1 $x+y=6$
$x-y=4$

2 $2x-y=5$
$x+y=7$

3 $3x-2y=2$
$x+2y=6$

4 $5x+y=9$
$2x+y=6$

5 $x-5y=2$
$x+3y=10$

6 $2x-3y=2$
$5x-3y=14$

7 $3x+5y=13$
$3x-y=1$

8 $4x-5y=8$
$x+5y=2$

9 $6x-y=9$
$2x-y=5$

10 $x-4y=2$
$x+3y=-5$

11 $4x-5y=-9$
$2x+5y=18$

12 $7x-3y=2$
$7x-y=-4$

13 $2x+7y=3$
$2x-3y=13$

14 $x-8y=-34$
$7x+8y=18$

15 $6x-4y=2$
$3x+4y=4$

1B

Solve each of the following pairs of equations to find values for x and y.

1 $4x-y=0$
$3x+2y=11$

2 $x+5y=17$
$2x-3y=-5$

3 $3x-7y=1$
$2x+y=12$

4 $x-2y=6$
$4x-3y=4$

5 $3x-4y=-23$
$5x+2y=5$

6 $5x-3y=11$
$4x-6y=7$

7 $3x-2y=8$
$2x-3y=12$

8 $2x-7y=0$
$3x-4y=13$

9 $5x+3y=9$
$2x-5y=16$

10 $3x-8y=11$
$4x+5y=-1$

11 $5x-3y=-11$
$7x+6y=-12$

12 $3x-7y=-1$
$7x-8y=31$

13 $2x-3y=-3$
$4x+5y=-6$

14 $5x-2y=-4$
$2x-3y=5$

15 $3x+4y=5$
$9x+2y=0$

1C

Solve the following:

1 $x+5y=0$
$2x+3y=-7$

2 $2x-3y=2$
$6x-5y=10$

3 $x-y=4$
$x+4y=-11$

4 $x+3y=-7$
$2x-3y=-5$

5 $5x-6y=1$
$3x-2y=1$

6 $6x-7y=7$
$9x-4y=4$

7 $2x+y=6$
$6x-y=12$

8 $2x-7y=22$
$3x+2y=8$

9 $3x-y=3$
$5x+2y=-17$

10 $6x-7y=5$
$5x-3y=7$

11 $x=3+y$
$x=11+3y$

12 $2x-11=3y$
$x+2y+12=0$

13 $5x=13+y$
$2y=8-7x$

14 $4x=3y-13$
$6x+5y+10=0$

15 $2x=y+1$
$4x=6y$

1D

1 On the same axes draw the lines $y=x+1$ and $y=7-x$. Find the co-ordinates of the point which lies on both lines. These give the solution to the pair of simultaneous equations $y=x+1$ and $y=7-x$. Check your answer by calculation or substitution.

2 Solve the simultaneous equations $y-3x=0$ and $2y-x=5$ by drawing the two lines on the same axes. Check your answer by calculation or substitution.

3 On the same axes draw the appropriate lines to solve the equations $4y=3x+1$ and $y=2x-1$. Check by calculation or substitution.

4 By drawing the lines $2y+x=8$, $y=x+1$ and $2y=x+1$ on the same axes find the solutions to three pairs of simultaneous equations. State the equations and their solutions.

5 Plot the points (2,3) (3,4) and (1,6). Draw a straight line to pass through two of these points and write down its equation. Repeat twice.

Using the equations you have found, form three pairs of simultaneous equations and check that the original points give the solutions to these simultaneous equations.

Solve graphically the following equations:

6 $x+2y=7$
$2x-y=4$

7 $3x+y=4$
$2x-y=6$

8 $3y=x-4$
$2y=x-2$

9 $x-y=6$
$x+2y=0$

10 $x+y=2$
$2x+3y=7$

1E

1 Alison has 4 more coloured pens than Brenda, and together they have 20. Let Alison have a pens and Brenda have b pens and write two equations in a and b. Use these to find how many coloured pens each girl has.

2 Sarah found that 2 cherry buns and 3 chocolate cakes would cost her 60p. She only had 50p and could just afford 4 cherry buns and 1 chocolate cake. Write two equations and use them to find the cost of each type of cake.

3 The sum of Gordon's age and his father's age is 39 years. Twice his father's age is 29 years more than five times Gordon's age. Let g and f represent their ages in years and write two equations relating g and f. Solve these and find how old each is.

4 There are two types of bricks in a construction set. Find the weight of one brick of each type knowing that 9 white and 5 red bricks together weigh 294 g, and that 6 white and 8 red bricks together weigh 336 g.

5 Martin is 8 years older than Neil. In one year's time Martin will be twice as old as Neil. How old are the boys now?

6 David is working out how to spend the pocket money he has saved up for his holiday. He finds that if he has 2 boat trips and 12 ice creams it will cost him £4.20, but if he has 3 boat trips he can only afford 4 ice creams for the same money. Find the cost of a boat trip and also of an ice cream.

7 5 pencils and 2 rubbers cost 55p, and 6 pencils and 1 rubber cost 52p. Find the cost of a rubber assuming that all the pencils cost the same and all the rubbers cost the same.

8 Mark is three years older than Louise and four times her age is the same as three times his age. Find their ages.

9 Three oranges and two lemons cost 50p altogether. If two oranges and three lemons cost 45p, find the price of one orange and of one lemon.

10 Mrs White's laundry bill is 96p for 2 sheets and 2 pillowcases. For 4 sheets and 3 pillowcases she has to pay £1.73. If she decides to wash the pillowcases herself, how much would she have to pay the laundry for doing 4 sheets?

11 On a rail excursion to the seaside, the fare for six adults and ten children was £19. The fare for two adults and seven children was £10. What was the fare for an adult, and what was it for a child?

12 A coach operator uses two types of coach, a Luxury Coach and a Cosy Coach. Four of the first and six of the second can together carry 524 passengers. Six of the first and four of the second can together carry 516 passengers. How many passengers can be carried in each kind of coach?

(*Hint* Your equations can be simplified by dividing them right through by 2.)

13 Peter has twenty marks more than John in a mathematics examination, but John has twice as many marks as Peter in a chemistry examination.

   a) If Peter's marks are m and c, what are John's marks?
   b) If in the two examinations Peter's marks add up to 100 and John's to 120, write down two equations in m and c.
   c) Solve these equations and find how many marks the two boys had in each examination.

14 In a school cricket match, Charles carried his bat, nearly making a century. Robert was almost run out, but scraped home and scored another ten runs before being clean bowled. If Robert had been run out, Charles's score would have been three times as big as Robert's, but if Robert had made six more runs than he actually did, Charles's score would have been just double Robert's score.

a) If Robert's score was r, what would his score have been if he had been run out as above?

b) What would his score have been if he had made six more runs than he actually made?

c) Taking Charles's score as c, write down two equations in c and r, and hence find Charles's final score.

15 Wendy's father is a circus manager. She and her brother Nigel are both good acrobats and sometimes perform in the ring. They parade around the ring, Wendy standing upright on Nigel's head, the two being dressed as a striking-looking young lady just 270 centimetres high. In another act Wendy, walking on stilts hidden beneath her dress, clowns as Nigel's wife. The stilts lift her 50 cm from the ground, and she is then just the same height as Nigel. How tall is Wendy and how tall is Nigel?

* 16 The number 27 contains two digits, 2 and 7. If the digits are reversed the number becomes 72.

a) Is it true to say that the number is $(10 \times 2) + 7$ and that the number with the digits reversed is $(10 \times 7) + 2$?

b) If the digits are 4 and 6, write down the number and the 'reverse' number in a similar way to a).

c) If the digits are x and y, what is the number, and what is the number formed by reversing the digits?

d) If the sum of the two numbers in c) is 99 and the difference is 27, write down two equations for x and y.

e) Solve these equations and hence find the two numbers.

* 17 If you subtract 18 from a two figure number, the digits are reversed. The sum of the two digits is 10. What are the two numbers?

* 18 If the digits of a number are reversed, the number is increased by 36. The two digits added together give a total of 12. What was the original number?

* 19 The line $y = mx + c$ goes through the points (1,5) and (4,11).

a) Substitute the first set of co-ordinates in $y = mx + c$ and get an equation connecting m and c.

b) Repeat a) for the second set of co-ordinates.

c) Solve the two equations and find m and c. What is the equation of the line? Check that it does go through the two given points.

* 20 In a similar way find the equation of the line that goes through (1,4) and (2,7).

* 21 Find also the equation of the line through the points (3,2) and (6,5).

2 Functions

2A Simple Functions

Note In this chapter all functions are defined in the rational domain unless otherwise stated.

1 The function *f* is defined in the domain of the positive integers and $f : x \rightarrow$ the largest prime factor of *x*. For the domain {6, 8, 10, 12, 14} complete the map.

6	
8	
10	→ 2
12	
14	

2 The function *F* is defined in the domain of the positive integers and $F : x \rightarrow$ the smallest prime factor of *x*. For the domain {4, 10, 15, 25, 39} complete the mapping diagram.

3 If *f* is defined in the integral domain 10–99 and $f : x \rightarrow$ the number obtained when the digits of *x* are added, complete the map for the domain {10, 11, 21, 26, 33}

4 If the function *F* is defined in the integral domain 10–99 as $F : x \rightarrow$ the number obtained by squaring and adding the digits of *x*, draw the mapping diagram for the domain {12, 23, 29, 32, 40}

5 If the function *g* is defined as $g : x \rightarrow$ the next highest integer above *x*, draw the map for the domain {1·5, 2·3, 2·7, 5, 6·2}

6 $f : x \rightarrow 2x - 1$ Draw a map for the domain {−2, −1, 0, 1, 2}

7 For each of the following functions, write the range for the domain {−3, −2, −1, 0, 1, 2}

   a) $f : x \rightarrow 3x + 2$ b) $f : x \rightarrow 5 - x$ c) $f : x \rightarrow \dfrac{x}{2}$

   d) $f : x \rightarrow 1 - 2x$ e) $f : x \rightarrow x^2$

8 If $f : x \rightarrow \dfrac{12}{x}$ find the range for the domain {1, 2, 3, 4, 6, 8}

9 If $F : x \rightarrow 2^x$ find the range for the domain {−1, 0, 1, 2, 3}

13

10 Find the range for the domain $\{-2, -1, 0, 1, 2\}$ for each of the following functions:

a) $f : x \rightarrow 4x - 2$ 　　　b) $f : x \rightarrow 7 - x$ 　　　c) $f : x \rightarrow \dfrac{x+5}{2}$

d) $f : x \rightarrow \dfrac{16}{x+3}$ 　　　e) $f : x \rightarrow 2x + 3$

11 If $f : x \rightarrow 3x - 5$ find the values of the following:

a) $f(2)$ 　　b) $f(-3)$ 　　c) $f(0)$ 　　d) $f(y)$ 　　e) k if $f(k) = 4$

12 If $f : x \rightarrow 6 - x$ find the values of the following:

a) $f(-3)$ 　　b) $f(-6)$ 　　c) $f(1)$ 　　d) $f(2y)$ 　　e) k if $f(k) = 8$

13 Given that $f : x \rightarrow 2x + 1$ and $g : x \rightarrow 3 - x$ find the values of:

a) $f(-2)$ 　　b) $g(1)$ 　　c) $fg(1)$ 　　d) $gf(3)$ 　　e) $fg(k)$ 　　f) $gf(n)$

14 Given that $f : x \rightarrow \dfrac{x}{2}$ and $g : x \rightarrow 4x - 1$ find the values of:

a) $f(-4)$ 　　b) $g(1)$ 　　c) $fg(1)$ 　　d) $f(8)$ 　　e) $gf(8)$ 　　f) $fg(x)$

15 If $F : x \rightarrow 6 - 2x$ and $f : x \rightarrow \dfrac{6}{x}$ find the values of:

a) $F(-1)$ 　　b) $fF(2)$ 　　c) $f(-3)$ 　　d) $Ff(2)$ 　　e) $Ff(12)$ 　　f) $Ff(x)$

2B Relationships and their Inverses

1 The Grant family consists of Angela, Brian, David, Catherine and Eliza, in that order with Angela being the eldest.
Complete the arrow diagram to show the relationship 'is older than'.
Draw a similar diagram, to show the relationship 'is younger than'.

2 Draw a similar diagram for the same family to show the relationship 'is the brother of'.
If the arrows in this diagram are reversed what relationship, if any, do they illustrate?

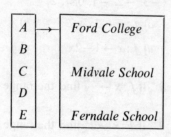

3 The two youngest children in the Grant family go to Ferndale School, the two boys to Midvale School, and Angela goes to Ford College.

Complete the diagram to show the relationship 'is a pupil at'. What relationship is shown if the arrows are all reversed?

A $\rightarrow$	Ford College
B	
C	Midvale School
D	
E	Ferndale School

4 Five Venture Scouts met at an International Camp. John came from England, Pierre from France, Andrew from Australia, Kurt from Sweden and Hans from Germany. Write the boys' names in one box and the countries in another, and put in arrows to show the relationship 'is from'. When the arrows are reversed the new relationship is the inverse of the original relationship. What is the inverse relationship in this case?

5 Put in arrows to show the relationship 'has symmetry of order'. What is the inverse relationship, represented by reversing all the arrows?

Equilateral triangle		2
Square		3
Rectangle		
Parallelogram		4
Regular hexagon		6

6 Put arrows into the diagram to show the relationship 'is a factor of'. What is the inverse relationship?

2	6
3	8
4	9
5	10

7 Complete the diagram to show the relationship 'is the highest prime factor of'. If the arrows are reversed, does the map represent the inverse relationship found in question 6? If your answer is 'no', give reasons.

2	6
3	8
5	9
	10

8 Complete the diagram to show the relationship 'passes through'. What is the inverse of this relationship, i.e. what is the relationship when all the arrows are reversed?

$y=x$	$(-1, 1)$
$y=x+2$	$(0, 0)$
$y=4-x$	$(1, -1)$
$y=3x-4$	$(1, 3)$
$y+x=0$	$(2, 2)$
	$(3, 5)$

9 The diagram shows a rhombus *ABCD* and its two diagonals. By putting the four sides into the domain and into the range, show the relationship 'is parallel to'. What is the inverse? For the same rhombus complete a mapping to show 'is perpendicular to'. What is the inverse? What do you notice about both the inverses in this question?

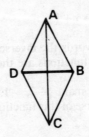

10 Complete the diagram to show the relationship 'divided into 12 gives'. What is the inverse of this relationship?

1	12
2	6
3	4
4	3
6	2
12	1

2C Functions and their Inverses

1 By definition a relationship is a function if each element in the domain maps on to one and only one member of the range.
Look at each of the following relationships and state whether it is a function or not.

   a) Pupils in the class → day on which their birthday falls in 1978.
   b) Points on a graph → their images after reflection in $x=y$.
   c) Mothers → their sons.
   d) Books → authors.
   e) Telephones → telephone numbers.

2 State whether the inverse of each of the relationships in question 1 is a function or not.

3 Complete the mapping diagram for the domain $\{-2, -1, 0, 1, 2\}$ and the function $f:x \to x+2.$
Write down the inverse of this function. Is the inverse itself a function?

4 Taking the same domain $\{-2, -1, 0, 1, 2\}$ complete a mapping for each of the following functions:

   a) $f:x \to 3x$ *b)* $f:x \to \dfrac{x}{2}$ *c)* $f:x \to x-3$

   d) $f:x \to 5-x$ *e)* $f:x \to \dfrac{6}{x}$ (exclude 0 from the domain)

Write the inverse of each and state whether it is a function or not.
Two of the above functions have a special property. Which are they? What is it?

5 The following functions are defined in the rational domain. Describe them in words.

   a) $f:x \to 4x$ *b)* $f:x \to \dfrac{x}{5}$ *c)* $f:x \to \dfrac{8}{x}$

   d) $f:x \to x+1$ *e)* $f:x \to x-5$ *f)* $f:x \to 7-x$

6 Write in words the inverse of each of the functions in question 5.
Which of the functions are their own inverses?

7 For the domain $\{-2, -1, 0, 1, 2\}$ complete the mapping of the function $f:x \to x^2$.
Find the inverse of this function. Is the inverse a function or not? Give reasons for your answer.

16

Note In questions 8 to 12 the functions are defined in the rational domain.

8 For the function f the inverse is written f^{-1}.
If $f:x \rightarrow x+6$ find the values of:

    a) $f(-3)$ b) $f(0)$ c) $f^{-1}(-2)$ d) $f^{-1}(0)$ e) $f^{-1}(6)$

9 If $f:x \rightarrow 12-x$ and $g:x \rightarrow x-3$ find the values of:

    a) $f(-2)$ b) $f^{-1}(4)$ c) $g(5)$ d) $g^{-1}(-1)$ e) $fg^{-1}(-1)$ f) $gf^{-1}(0)$

10 If $F:x \rightarrow \dfrac{15}{x}$ and $f:x \rightarrow 3x$ find the values of:

    a) $F(-3)$ b) $F^{-1}(5)$ c) $f^{-1}(6)$ d) $Ff(x)$ e) $(Ff)^{-1}(x)$ f) $Ff^{-1}(x)$

11 If $F:x \rightarrow \dfrac{x}{4}$ and $f:x \rightarrow x-8$ find the values of:

    a) $F^{-1}(x)$ b) $f^{-1}(x)$ c) $FF^{-1}(x)$ d) $fF^{-1}(1)$ e) $F^{-1}f(6)$ f) $f^{-1}f(-3)$

12 If $f:x \rightarrow 5-x$ and $g:x \rightarrow \dfrac{8}{x}$ find:

    a) a if $f(a)=3$ b) b if $f^{-1}(b)=6$ c) c if $g(c)=2$ d) d if $g(d)=-2$
    e) e if $fg(e)=-3$

2D Flowcharts

Note In this section all functions are defined in the rational domain unless otherwise stated.

1 The flowchart shows the function $f:x \rightarrow 2x-5$.

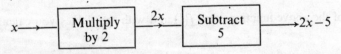

The reverse of this will give the inverse of the function.

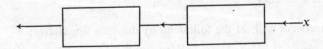

Write in the boxes the inverses of the two functions which together make up the original function, and hence find the inverse.

    For the domain $\{0, 1, 2, 3, 4\}$ complete the mapping for $f:x \rightarrow 2x-5$. With all the arrows reversed the mapping should then be true for the inverse you have just found from the flowchart. Check that this is so.

2 This flowchart shows the function $f:x \to 8-3x$.

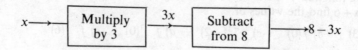

By putting in the boxes the two separate inverses, find the inverse of the original function.

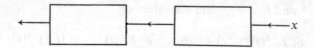

For the domain $\{-2, -1, 0, 1, 2\}$ complete the mapping for $f:x \to 8-3x$. Using the inverse which you have just found, check that it is true for these values.

3 Draw a flowchart for $f:x \to 3x+1$ and a second one to find its inverse. Is the inverse a function?
Taking $\{-2, -1, 0, 1, 2\}$ as the domain, complete the mapping for $f:x \to 3x+1$ and check the inverse.

4 Write down the inverse of each of these functions. You will find it helpful to draw a flowchart for each.

a) $f:x \to 4x-3$ b) $f:x \to \frac{1}{2}(x+1)$ c) $f:x \to 7-2x$
d) $f:x \to 5+2x$ e) $f:x \to \frac{1}{2}(6-x)$

5 Draw a flowchart for $f:x \to \dfrac{6}{x+2}$ and a second one to find its inverse.

Taking $\{-1, 0, 1, 2\}$ as the domain, complete the mapping for the function and check that the inverse which you have found is correct.

6 Find the inverse of each of these functions by drawing flowcharts:

a) $f:x \to \dfrac{8}{x-2}$ b) $f:x \to \dfrac{5+x}{2}$ c) $f:x \to \dfrac{6}{3-x}$

d) $f:x \to \frac{1}{4}(10-x)$ e) $f:x \to 4-\dfrac{x}{2}$

For each of the above functions find $f(1)$ and with the values you calculate check your inverses.

7 Draw a flowchart for $f:x \to 3x^2$ and a second one for its inverse. Complete a mapping for this function taking $\{-2, -1, 0, 1, 2\}$ as the domain. Check the inverse which you have found. Is it a function?

8 Find the inverse of each of the following by drawing flowcharts:

a) $f:x \to 12-5x$ b) $f:x \to \frac{1}{2}(4x+3)$ c) $f:x \to \frac{1}{2}(5-2x)$

d) $f:x \to 8-\dfrac{5}{x}$ e) $f:x \to \dfrac{12}{2x+1}$

9 Find the inverse of each of the following by drawing flowcharts:

a) $f:x \to 2x^2$ b) $f:x \to x^2+2$ c) $f:x \to \dfrac{x^2}{4}$ d) $f:x \to \dfrac{4}{x^2}$

State whether the inverses you have found are functions or not.

10 Find the inverse of each of the following:

a) $f:x \to \dfrac{6+2x}{5}$ b) $f:x \to 5-3(x+1)$ c) $f:x \to \dfrac{15}{9-2x}$

d) $f:x \to 2-\dfrac{4}{x}$

Calculate $f(2)$ for each of the functions and with the values you find check the inverse of each.

2E Miscellaneous

Note In this section all functions are defined in the rational domain except where otherwise stated.

1 If $f:x \to 2x+6$ and $g:x \to 1-x$, calculate:

a) $f(-3)$ b) $g(-3)$ c) $g^{-1}(2)$ d) $fg^{-1}(2)$ e) $f^{-1}(9)$

2 If $f:x \to \frac{1}{2}(x+3)$ and $g:x \to \dfrac{x}{2}$ find:

a) $f(1)$ b) $g^{-1}(-2)$ c) $fg^{-1}(0)$ d) $f^{-1}(x)$ e) y if $f^{-1}(y)=3$

3 If $f:x \to \dfrac{9}{x+1}$ and $g:x \to 4-x$ calculate:

a) $f(2)$ b) $g(2)$ c) $fg(2)$ d) $gf(2)$ e) $fg(x)$

4 If $f:x \to \dfrac{6}{x-2}$ find f^{-1}.

Calculate $f(-1)$ $f(5)$ $f^{-1}(6)$ $ff^{-1}(6)$

5 If $F:x \to 4-2x$ find F^{-1}.

Calculate $F(0)$ $F^{-1}(4)$ $FF(4)$ $F^{-1}F(0)$

6 If $F:x \to \frac{1}{2}(5-x)$ find F^{-1}.

Calculate $F(1)$ $FF(1)$ $F^{-1}(2)$ $FF^{-1}(x)$

7 If $F:x \to 10-3x$ and $f:x \to \dfrac{6}{x}$ calculate:

a when $F(a)=2a$, b when $F^{-1}(b)=-2$,
c when $f^{-1}(c)=-1$, d when $Ff(d)=8$

8 If $f:x \to 3x+2$ and $g(x)=x^2$ find:

a) $f(-1)$ b) $g^{-1}(9)$ c) $fg(x)$ d) y if $fg(y)=14$ e) $g^{-1}f(x)$

9 If $f:x \to (x-2)(x+1)$ complete the mapping for the domain $\{-2, -1, 0, 1, 2\}$

10 If $f:x \to 2^{x-1}$ complete the mapping for the domain $\{-1, 0, 1, 2, 3\}$. Find x if $f(x)=32$.

11 If $f:x \to x-2$ and $g:x \to \dfrac{x}{2}$ find:

a) x such that $f(x)=g(x)$ b) x such that $ff(x)=g(x)$
c) x such that $gg(x)=f(x)$ d) x such that $g^{-1}(x)=f^{-1}(x)$

12 If $F:x \to$ the smallest integer which is equal to or greater than x, find:

a) $F(2\cdot6)$ b) $F(-1\cdot5)$ c) $F(\frac{5}{2})$ d) $\frac{1}{2}F(5)$

13 If the domain of the function G is the positive integers and $G:x \to$ the sum of the prime factors of x (excluding 1 and x), calculate:

a) $G(6)$ b) $G(10)$ c) $G(12)$ d) $G(15)$ e) $G(21)$

Write down two other possible values of x such that $G(x)=G(6)$.

14 If the domain of the function f is the positive integers and $f:x \to$ the sum of all the positive integers up to and including x, calculate:

a) $f(2)$ b) $f(5)$ c) $f(8)$ d) $f(n)$

15 The functions F and f are defined as follows in the domain of the positive integers:
$F:x \to$ the highest factor of x (other than x)
$f:x \to$ the highest prime factor of x (other than x).

Find a) $F(6)$ b) $f(6)$ c) $F(12)$ d) $f(12)$ e) $F(15)$ f) $f(15)$
g) $F(20)$ h) $f(20)$

If $F(x)=f(x)$ what can you say about x?

16 If $f:x \to ax+b$ and $f(1)=7$ and $f(-1)=3$ find values for a and b. Write down $f^{-1}(x)$.

17 If $f:x \to ax+b$ and $f(x)=f^{-1}(x)$ what do you know about the values of a and b if they are neither 0 nor 1?

18 If $f:x \to ax+b$ and $f(1)=8$ what is the value of $f^{-1}(8)$? If $f(k)=l$ what is $f^{-1}(l)$?

19 If $f:x \to \dfrac{a}{x}$ what is $f^{-1}(x)$?
$f(5)=3$. What is the value of a?

20 If $f:x \to a-bx$ and $f(-1)=7$ and $f(1)=5$, find the values of a and b.
What is $f^{-1}(7)$?

3 Logarithms of Numbers Less than 1

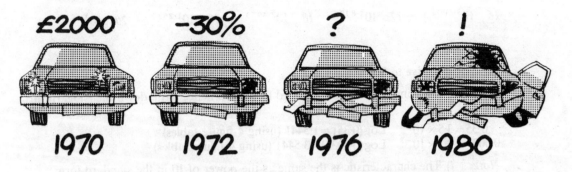

£2000   −30%   ?   !
1970   1972   1976   1980

Note   In this chapter, where rounding off is necessary, round fives upwards.

3A

Write each of these numbers in standard form, i.e. $a \times 10^b$ where a lies between 1 and 10, and b is an integer.

1 2 500 000	*2* 3750	*3* 106 000	*4* 0·027	*5* 0·0003
6 0·004 55	*7* 0·000 004 8	*8* 0·035	*9* 0·000 012	*10* 0·000 008 75

3B   (See 3B′ for an alternative approach)

Write each of these numbers as a power of 10 using standard form and logarithms, e.g.

$$0·025 = 2·5 \times 10^{-2} = 10^{0·398} \times 10^{-2} = 10^{\bar{2}·398} \text{ (using 3-figure tables)}$$

or $10^{\bar{2}·3979}$ (using 4-figure tables)

Note   Numbers 11 to 20 will need rounding off if 3-figure tables are used.

1 0·750	*2* 0·005 75	*3* 0·087	*4* 0·000 456
5 0·285	*6* 0·0213	*7* 0·000 019 2	*8* 0·517
9 0·005 63	*10* 0·003 79	*11* 0·007 234	*12* 0·081 25
13 0·000 617 2	*14* 0·2189	*15* 0·003 651	*16* 0·004 568
17 0·000 018	*18* 0·007 963	*19* 0·000 121 4	*20* 0·000 003 787

3C   (See 3C′ for an alternative approach)

Using the reverse process, write down the numbers to which the following are equivalent.

Note   Numbers 11 to 20 will need rounding off if 3-figure tables are used.

1 $10^{\bar{1}·751}$	*2* $10^{\bar{1}·523}$	*3* $10^{\bar{2}·756}$	*4* $10^{\bar{1}·988}$	*5* $10^{\bar{4}·627}$

6 $10^{\overline{3}\cdot122}$ 7 $10^{\overline{3}\cdot525}$ 8 $10^{\overline{2}\cdot129}$ 9 $10^{\overline{2}\cdot879}$ 10 $10^{\overline{5}\cdot464}$

11 $10^{\overline{2}\cdot1668}$ 12 $10^{\overline{1}\cdot2939}$ 13 $10^{\overline{4}\cdot2684}$ 14 $10^{\overline{2}\cdot2691}$ 15 $10^{\overline{2}\cdot9718}$

16 $10^{\overline{3}\cdot1002}$ 17 $10^{\overline{1}\cdot8774}$ 18 $10^{\overline{2}\cdot4953}$ 19 $10^{\overline{1}\cdot2637}$ 20 $10^{\overline{3}\cdot1819}$

3B′ (Alternative to 3B)

Write the following numbers in standard form, and hence write down their logarithms, eg.

$0{\cdot}35 = 3{\cdot}5 \times 10^{-1}$ $\text{Log}(0{\cdot}35) = \overline{1}{\cdot}5441$ (using 4-figure tables)
$0{\cdot}0035 = 3{\cdot}5 \times 10^{-3}$ $\text{Log}(0{\cdot}0035) = \overline{3}{\cdot}5441$ (using 4-figure tables)

Note   i) The characteristic is the same as the power of 10 in the standard form.
   ii) Only the characteristic is negative. The mantissa is always positive.
   iii) Nos. 11 to 20 will need rounding off if 3-figure tables are used.

1 0·750   2 0·005 75   3 0·087   4 0·000 456   5 0·285

6 0·0213   7 0·000 001 92   8 0·517   9 0·005 63   10 0·003 79

11 0·007 234   12 0·081 25   13 0·000 617 2   14 0·2189   15 0·003 651

16 0·004 568   17 0·000 018   18 0·007 963   19 0·000 121 4   20 0·000 003 787

3C′ (Alternative to 3C)

You are given the logarithms of twenty numbers. Write down these numbers, first in standard form and then in ordinary form, e.g.

Logarithm 2·5469   Number $3{\cdot}523 \times 10^{-2}$ or 0·03523

Note   Numbers 11 to 20 will need rounding off if 3-figure tables are used.

1 $\overline{1}{\cdot}751$ 2 $\overline{1}{\cdot}523$ 3 $\overline{2}{\cdot}756$ 4 1·988   5 $\overline{4}{\cdot}627$

6 $\overline{3}{\cdot}122$ 7 $\overline{3}{\cdot}525$ 8 $\overline{2}{\cdot}129$ 9 $\overline{2}{\cdot}879$ 10 $\overline{5}{\cdot}464$

11 $\overline{2}{\cdot}1668$ 12 $\overline{1}{\cdot}2939$ 13 $\overline{4}{\cdot}2684$ 14 $\overline{2}{\cdot}2691$ 15 $\overline{2}{\cdot}9718$

16 $\overline{3}{\cdot}1002$ 17 $\overline{1}{\cdot}8774$ 18 $\overline{2}{\cdot}4953$ 19 $\overline{1}{\cdot}2637$ 20 $\overline{3}{\cdot}1819$

3D

Remembering that the number written $\overline{1}{\cdot}2576$ means $-1 + 0{\cdot}2576$ and has to be used in calculations with care, work out the sums of each of the following pairs of numbers.

1	2	3	4	5
$\overline{2}{\cdot}3$	$\overline{1}{\cdot}8$	$1{\cdot}3$	$\overline{2}{\cdot}8$	$\overline{2}{\cdot}9$
$\overline{1}{\cdot}2$	$\overline{1}{\cdot}7$	$\overline{2}{\cdot}2$	$0{\cdot}3$	$\overline{1}{\cdot}6$

6	7	8	9	10
$\overline{3}{\cdot}7$	$\overline{2}{\cdot}8$	$3{\cdot}1$	$\overline{3}{\cdot}1$	$2{\cdot}4$
$1{\cdot}5$	$1{\cdot}9$	$\overline{2}{\cdot}8$	$\overline{1}{\cdot}8$	$\overline{1}{\cdot}7$

  1·8   **12**   1·4   **13** $\bar{2}$·1   **14** $\bar{4}$·2   **15** $\bar{1}$·5

11 1·8	**12** 1·4	**13** $\bar{2}$·1	**14** $\bar{4}$·2	**15** $\bar{1}$·5
$\bar{1}$·6	2·7	$\bar{1}$·3	0·7	3·8

Numbers 16–30: Subtract the second number from the first in each of numbers 1–15.

Note   Check your subtractions by adding the two bottom lines. Their sum should give the top line.

3E

Subtract the second number from the first in each of the following:

1 1·5	**2** 1·2	**3** $\bar{2}$·7	**4** 0·3	**5** $\bar{1}$·2
1·2	1·6	$\bar{1}$·9	1·5	2·1

6 2·6	**7** $\bar{2}$·3	**8** 1·8	**9** 3·2	**10** 3·5
1·2	$\bar{3}$·5	2·6	$\bar{1}$·6	2·7

11 3·4	**12** 2·5	**13** $\bar{2}$·4	**14** $\bar{3}$·1	**15** $\bar{1}$·6
2·2	$\bar{2}$·8	3·5	2·7	$\bar{3}$·9

Note   Check your subtractions by adding the two bottom lines. Their sum should give the top line.

3F

Work out the following:

1 $\bar{1}$·5 × 2	**2** $\bar{2}$·3 × 2	**3** $\bar{1}$·7 × 3	**4** $\bar{1}$·4 × 3
5 $\bar{2}$·9 × 3	**6** $\bar{2}$·6 ÷ 2	**7** $\bar{2}$·8 ÷ 2	**8** $\bar{3}$·9 ÷ 3
9 $\bar{1}$·8 ÷ 2	**10** $\bar{1}$·1 ÷ 3	**11** $\bar{1}$·2 ÷ 2	**12** $\bar{2}$·5 ÷ 3
13 $\bar{1}$·4 ÷ 3	**14** $\bar{3}$·6 ÷ 2	**15** $\bar{2}$·8 ÷ 3	**16** $\bar{3}$·2 ÷ 2
17 $\bar{2}$·4 ÷ 4	**18** $\bar{4}$·7 ÷ 3	**19** $\bar{3}$·5 ÷ 2	**20** $\bar{1}$·3 ÷ 3

3G

Work out the value of each of the following using logarithms:
(Note that 3-figure tables can be used for all these questions without preliminary correction.)

1 2·53 × 0·716	**2** 0·158 × 0·362	**3** 3·92 × 0·216
4 0·009 78 × 0·532	**5** 0·266 × 0·0397	**6** 0·0885 × 0·103
7 0·0517²	**8** 0·234³	**9** 0·535 ÷ 0·0172
10 0·696 ÷ 0·0928	**11** 0·0426 ÷ 0·735	**12** 1·43 ÷ 1·98
13 0·532 ÷ 0·007 61	**14** 29·6 ÷ 0·448	**15** 0·723 ÷ 0·896

3H

Work out the value of each of the following using logarithms:

1 $2 \cdot 7 \times 0 \cdot 3125$ **2** $1 \cdot 85 \times 0 \cdot 7637$ **3** $0 \cdot 0296 \times 5 \cdot 3$

4 $0 \cdot 01691 \times 0 \cdot 8372$ **5** $12 \cdot 4 \times 0 \cdot 6514$ **6** $1 \cdot 83 \times 0 \cdot 07562$

7 $0 \cdot 00513 \times 0 \cdot 7271$ **8** $0 \cdot 09135 \times 0 \cdot 0719$ **9** $0 \cdot 09632^2$

10 $0 \cdot 8573^3$ **11** $2 \cdot 56 \div 0 \cdot 1736$ **12** $0 \cdot 8726 \div 0 \cdot 8915$

13 $0 \cdot 0036 \div 0 \cdot 537$ **14** $0 \cdot 08311 \div 0 \cdot 003682$ **15** $1 \cdot 37 \div 5 \cdot 866$

16 $279 \cdot 5 \div 4220$ **17** $1 \cdot 069 \div 0 \cdot 03192$ **18** $0 \cdot 0376 \div 143$

19 $29 \cdot 43 \div 0 \cdot 324$ **20** $3420 \div 17600$

3I

Use logarithms to find the square roots of the following:

1 $0 \cdot 04272$ **2** $0 \cdot 4272$ **3** $0 \cdot 05639$ **4** $0 \cdot 5639$

5 $0 \cdot 000789$ **6** $0 \cdot 753$ **7** $0 \cdot 00658$

Use logarithms to find the cube roots of the following:

8 $0 \cdot 009262$ **9** $0 \cdot 09262$ **10** $0 \cdot 9262$ **11** $0 \cdot 00754$

12 $0 \cdot 754$ **13** $0 \cdot 0005573$ **14** $0 \cdot 469$

3J

Use logarithms to find the value of each of the following:

1 $\dfrac{0 \cdot 098 \times 0 \cdot 76}{4 \cdot 55}$ **2** $\dfrac{0 \cdot 923 \times 1 \cdot 78}{25 \cdot 6}$ **3** $\sqrt{\dfrac{0 \cdot 482}{0 \cdot 861}}$

4 $(0 \cdot 026)^2 \times 86 \cdot 5$ **5** $\sqrt{0 \cdot 482 \times 0 \cdot 024}$ **6** $\dfrac{0 \cdot 459 \times 1 \cdot 89}{0 \cdot 0813}$

7 $\dfrac{0 \cdot 00558}{0 \cdot 361 \times 0 \cdot 0819}$ **8** $\sqrt[3]{\dfrac{4 \cdot 63}{0 \cdot 852}}$ **9** $0 \cdot 637 \times \sqrt{\dfrac{14 \cdot 8}{25 \cdot 3}}$

10 $\sqrt[3]{0 \cdot 862} + \sqrt{0 \cdot 539}$ **11** $\sqrt{\dfrac{4 \cdot 17}{0 \cdot 842}}$ **12** $\sqrt[3]{\dfrac{0 \cdot 0361}{0 \cdot 184}}$

13 $\sqrt[3]{\dfrac{0 \cdot 146}{0 \cdot 00472}}$ **14** $\sqrt{\dfrac{8 \cdot 23}{9 \cdot 78}} + \sqrt{\dfrac{0 \cdot 365}{0 \cdot 942}}$ **15** $\sqrt[3]{\dfrac{96 \cdot 2}{124}} + \sqrt{\dfrac{0 \cdot 0375}{0 \cdot 0075}}$

16 $0 \cdot 3625 \times 5 \cdot 98$ **17** $\sqrt{0 \cdot 004235}$ **18** $0 \cdot 5836^3$

19 $\sqrt[3]{0 \cdot 07517}$ **20** $\dfrac{2 \cdot 364}{0 \cdot 7519}$ **21** $\dfrac{0 \cdot 02918 \times 0 \cdot 35}{1 \cdot 629}$

22 $\sqrt[3]{\dfrac{4\cdot176}{29\cdot82}}$

23 $\dfrac{10\cdot65\times0\cdot8155}{0\cdot04327}$

24 $\sqrt[3]{\dfrac{825\cdot6}{97\cdot72}}$

25 $0\cdot1632^2+\sqrt{0\cdot1632}$

3K

Work out the following using logarithms:

1 $12\cdot64\times0\cdot00729$

2 $1\cdot653\times0\cdot8912\times0\cdot00367$

3 $296\div0\cdot8522$

4 $47\cdot51\div312\cdot8$

5 $0\cdot0296\div0\cdot007153$

6 $\dfrac{0\cdot08333\times14\cdot7}{0\cdot6719}$

7 $\dfrac{2\cdot867\times1\cdot365}{52\cdot8}$

8 $0\cdot6891^2$

9 $0\cdot743^3$

10 $\sqrt{0\cdot06815}$

11 $\sqrt{0\cdot0097}$

12 $\sqrt[3]{0\cdot2168}$

13 $\sqrt{0\cdot8979}$

14 $\sqrt[3]{0\cdot07156}$

15 $\sqrt{\dfrac{2\cdot56}{9\cdot81}}$

16 $\sqrt{\dfrac{0\cdot01926}{0\cdot5831}}$

17 $\dfrac{0\cdot02462\times0\cdot983}{0\cdot5466}$

18 $\dfrac{0\cdot8491^2}{0\cdot07128}$

19 $1\cdot536\times\sqrt{0\cdot8614}$

20 $\sqrt[3]{\dfrac{1\cdot843}{0\cdot4724}}$

3L

Work out the following using logarithms:

1 $0\cdot003793\times0\cdot5361$

2 $0\cdot8143^2\times0\cdot08379$

3 $\sqrt{\dfrac{0\cdot5286}{0\cdot06381}}$

4 $2\cdot95\times\sqrt{\dfrac{6\cdot43}{16\cdot8}}$

5 $0\cdot6972^2+\sqrt{0\cdot1026}$

6 $\sqrt{0\cdot0875}+\sqrt[3]{0\cdot6723}$

7 $\dfrac{12\cdot57\times\sqrt{0\cdot003664}}{0\cdot8381}$

8 $\left(\dfrac{0\cdot1833}{0\cdot4562}\right)^2$

9 $\sqrt[3]{0\cdot128}+0\cdot2145^2$

10 $0\cdot07816^3$

11 $\sqrt[3]{\dfrac{1\cdot682}{0\cdot5144}}$

12 $0\cdot458\times\sqrt{\dfrac{0\cdot0891}{0\cdot9774}}$

13 $0\cdot637\times\sqrt{\dfrac{14\cdot8}{25\cdot3}}$

14 $\dfrac{0\cdot00558}{0\cdot361\times0\cdot0819}$

15 $\sqrt[3]{0\cdot862}\times\sqrt{0\cdot5391}$

16 $\sqrt{\dfrac{4\cdot17}{0\cdot8422}}$

17 $\sqrt[3]{\dfrac{96\cdot21}{124\cdot3}}+\sqrt{\dfrac{0\cdot03752}{0\cdot00749}}$

18 $\sqrt[3]{\dfrac{825\cdot6}{97\cdot72}}$

19 $\left(\dfrac{0.2356}{0.8192}\right)^2 \times 1.478$ **20** $0.5342^2 + \sqrt{0.05342}$

3M

Use logarithms to work out the answers to each of the following:

1 Find the volume of wood in a table top which measures $3.15\,\text{m} \times 1.25\,\text{m} \times 3.25\,\text{cm}$. Give your answer in cubic metres correct to 2 significant figures.

2 Find the volume of a cube of side $2.75\,\text{cm}$. Give your answer in *a*) cubic centimetres, *b*) cubic metres. What would be the volume of 1 million such cubes?

3 How many boxes each with a capacity of $0.0625\,\text{m}^3$ would be needed to give a total capacity of $1.5\,\text{m}^3$?

4 What is the length of the side of a square of area $0.275\,\text{m}^2$? Give your answer in *a*) metres, *b*) centimetres, both correct to 2 significant figures.

5 What is the length of the side of a cube of volume $0.082\,\text{m}^3$?

6 Water is poured from a jug holding 0.725 litres into a beaker which holds 0.18 litres. How many times can the beaker be filled without refilling the jug?

7 500 solid metal blocks each with a volume of $0.035\,\text{m}^3$ are melted down and reformed into 655 cubes all of the same size. Find the volume of one of these cubes.

8 $0.00813\,\text{m}^3$ of wood is made into 125 small cubes all identical in size. Find the length of the sides of these cubes correct to 2 significant figures.

9 $2.15\,\text{mm}$ of rain falls in 24 hours. What volume of water falls on a garage roof which measures $3.5\,\text{m}$ by $6.75\,\text{m}$?

10 A floor with an area of $24.01\,\text{m}^2$ is tiled completely by 400 square tiles. Find the area of one tile and the length of the sides of each tile.

11 A window is double glazed and consists of six panes, each $51\,\text{cm}$ by $63\,\text{cm}$. What is the total area of glass? Give your answer in square metres.

12 A rectangular field measures 243 metres along its shorter side and 315 metres along the other side. What is its area in hectares?

13 The base of a rain water tank measures $122\,\text{cm}$ by $153\,\text{cm}$ and it is $122\,\text{cm}$ high. How much water can it hold? Give your answer to 3 significant figures, *a*) in cubic metres *b*) in litres.

14 How deep is the water in question 13 when the tank contains 1600 litres of water? (Give your answer to the nearest centimetre.)

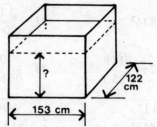

15 613 children in a school each pay 57p to go on an outing. What is the total sum they pay, correct to the nearest £?

16 A motor car costs £3370. At the end of a year its value has depreciated by 23%. What is its value, correct to the nearest £10?

17 At the end of a second year the value of the motor car in question 16 has depreciated by 19% of its value at the beginning of the second year. What is its value now (to the nearest £10)?

18 The average weight of 173 children is 43·2 kg. What is their total weight in tonnes, correct to 3 significant figures?

19 The total population of a town is 104 321. Approximately 27% of the total population work for one or other of the nationalised industries. How many altogether work for the nationalised industries? (Give your answer to the nearest hundred.)

20 There are 16 200 children of school age in the town mentioned in question 19. What percentage of the population is this? Give your answer correct to 2 significant figures.

21 Norman, the milk roundsman, works Saturdays and Sundays but has one day off every week, and three weeks holiday in the year.

a) Taking a year as 52 weeks, how many days does he work in a year?
b) If the total distance he travels on his round in a year is about 10 000 km, what is the average distance he travels every working day?

4 Changing the Subject of a Formula

4A

Rearrange each of the following equations to make x the subject of a simple formula.

1 $y=x+2$ **2** $y=x-5$ **3** $y=7-x$ **4** $3y=x-5$ **5** $4y=x+1$

6 $2y=1-x$ **7** $2y+x=3$ **8** $y-x=9$ **9** $y=3x$ **10** $y=\frac{1}{4}x$

11 $3y=5x$ **12** $y=\frac{2}{3}x$ **13** $y=2x+3$ **14** $y=2x-1$ **15** $y=3x-5$

16 $y=8-2x$ **17** $y+3x=5$ **18** $2y=3x+1$ **19** $3y+2x=7$ **20** $y=\frac{1}{2}(x+1)$

4B

Rearrange each of the following formulae to make the given letter the subject.

1 $C=2\pi r$ (r) **2** $S=\pi rl$ (l) **3** $I=\dfrac{PTR}{100}$ (R)

4 $V=u+at$ (t) **5** $A=\frac{1}{2}bh$ (h) **6** $I=m(v-u)$ (v)

7 $T=\dfrac{c}{l}x$ (l) **8** $V-v=-eu$ (v) **9** $mg=ma+T$ (a)

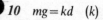 **10** $mg=kd$ (k)

4C

1 If $I=70$, $T=4$ and $R=5$, use the formula $I=\dfrac{PTR}{100}$ to find P.

2 If $v=10$, $u=4$ and $t=1\cdot5$, find the value of a in the formula $v=u+at$.

3 If $I = 24$, $m = 1.5$ and $v = 18$, use the formula $I = m(v - u)$ to find u.

4 If $A = 28$, $b = 5.6$ find the value of h in the formula $A = \frac{1}{2}bh$.

5 $A = \frac{1}{2}(a + b)h$ is the formula for finding the area of a trapezium with parallel sides of lengths a and b, h being the perpendicular distance between these sides.
When $a = 10$ cm, $b = 7$ cm and $h = 5$ cm, find the value of A.
Rearrange the formula and make a the subject.
Find the value of a if $A = 36$ cm^2, $b = 11$ cm and $h = 4$ cm.

6 The volume of a pyramid is given by the formula $V = \frac{1}{3}Ah$ where h is the perpendicular height and A is the area of the base. Find the volume of a pyramid of height 7 cm standing on a square base of side 9 cm.
Rearrange the formula to make h the subject. Find the value of h when $V = 50$ cm^3 and $A = 24$ cm^2.

7 Using the formula given in question 6, find the value of A given that $V = 192$ cm^3 and $h = 9$ cm.
If the pyramid stands on a square base, what is the length of the side of the square?

8 Using the formula $s = \frac{1}{2}(u + v)t$, find the value of s when $u = 2$, $v = 15$ and $t = 8$.
Rearrange the formula to make t the subject.
Find the value of t if $u = 5$, $v = 35$ and $s = 144$.

9 Rearrange the formula given in question 8 to make v the subject. If $s = 250$, $u = 4$ and $t = 15$, find the value of v.

10 Rearrange the formula $Pt = M(V - U)$ to make M the subject. Find the value of M if $P = 125$, $t = 0.2$, $V = 15$ and $U = 7$. Rearrange the same formula to make V the subject, and find the value of V if $P = 276$, $t = 0.25$, $M = 12$ and $U = 1.5$.

4D

Rearrange each of the following equations to express x in terms of the other letters.

1 $y = x + a$ **2** $y = x - b$ **3** $y = c - x$ **4** $ay = x - b$

5 $ay = c - x$ **6** $ay = bx + c$ **7** $y = m - nx$ **8** $ly + mx = n$

9 $y = \frac{p}{q}x$ **10** $y = x(a + b)$ **11** $y = \frac{x}{1 + m}$ **12** $\frac{y}{a} = \frac{x}{b}$

13 $\frac{l}{y} = \frac{m}{x}$ **14** $x^2 = y$ **15** $x^2 = a^2y$ **16** $l^2x^2 = my$

17 $y = p\sqrt{x}$ **18** $y = \sqrt{qx}$ **19** $y = \frac{a}{b}x^2$ **20** $y = \frac{1}{p}\sqrt{x}$

4E

Rearrange each of the following formulae to make the given letter the subject.

1 $A = \pi r^2$ (r) **2** $E = \frac{1}{2}mv^2$ (v) **3** $S = 4\pi r^2$ (r)

4 $V = \frac{1}{3}\pi r^2 h$ (h) (r)   **5** $V = \frac{4}{3}\pi r^3$ (r)   **6** $s = \frac{1}{2}gt^2$ (t)

7 $T = 2\pi\sqrt{\dfrac{l}{g}}$ (l) (g)   **8** $v^2 = u^2 + 2as$ (a)   **9** $S = 2\pi r(h+r)$ (h)

10 $I = m\left(\dfrac{a^2}{3} + d^2\right)$ (d) (a)

4F

1 Rearrange the formula $s = ut + \frac{1}{2}at^2$ to find u.
If $s = 55$, $a = 3\cdot5$ and $t = 4$, find the value of u.

2 Rearrange the formula $v^2 = u^2 + 2as$, to make u the subject.
Find the value of u if $v = 8$, $a = 2$ and $s = 7$.

3 If $v = 9$, $u = 5$ and $g = 10$, find the value of h from the equation $\dfrac{v^2}{2} = \dfrac{u^2}{2} + gh$.

4 From the equation $y = \dfrac{1}{p}\sqrt{x}$, find x if $y = 3$ and $p = 4$.

5 The volume of a cube of side a is given by the formula $V = a^3$.
The total mass of a solid cube of the same dimensions is given by $M = ma^3$ where m is the mass per unit volume.
Rearrange this formula to make a the subject.
Find the value of a if $M = 384\,g$ and $m = 6\,g/cm^3$.

6 a) Solve the equation $7x = 3x + 2$.
b) Rearrange the equation $ax = bx + c$ to find x in terms of a, b and c.

7 $y = 2\sqrt{x}$. If $x = 25$ find the value of y.

$y = \sqrt{2x}$. If $x = 25$ find the value of y.

Rearrange both of these equations to find x in terms of y.

8 Using the approximate formula $T = 6\sqrt{\dfrac{L}{10}}$, find the value of T when $L = 0\cdot9$.

Rearrange the formula and make L the subject. Find L when $T = 1\cdot5$.

9 Using the equation $y = \dfrac{x}{3} + \dfrac{x}{4}$, find the value of y when $x = 6$. Rearrange the equation and find x in terms of y.

10 Using the formula $I = \dfrac{ma^2}{2} + mx^2$ find the value of x if $I = 102\cdot5$ when $m = 10$ and $a = 4$.

5 Probability

5A

1 If you toss a coin, you are (theoretically) just as likely to throw a head as a tail. This means that in 100 tosses, you should (theoretically) get 50 heads and 50 tails. Toss a coin 100 times and see how close you get to this result.

If you work together in a group, the total number of tosses can be increased considerably. As the number of tosses is increased, are you more likely to get nearer to the theoretical answer, i.e. half the number of tosses being heads and half tails?

2 In question 1, the probability of being right if you call 'heads' is 1 in 2 or $\frac{1}{2}$. What is the probability of scoring a six if you throw one die?

Form a group and throw dice a large number of times. Find what proportion of your throws gives sixes.

Why do you need to make a large number of throws?

3 The table shows the possible total scores from throwing two dice. There are 36 ways of scoring. Copy and complete the table and use it to answer the following questions.

	1	2	3	4	5	6
1	2	3	4			
2			5			
3			6			
4						
5						
6				9	10	11	12

a) What is the least possible score and how many times does it occur in the table?
b) What is the maximum possible score and what is the probability of getting this?
c) What score appears most frequently in the table?
d) What is the probability of getting this score?
e) What is the probability of scoring an odd number?

4 Using the table from question 3, list the probabilities of scoring each total from 2 to 12.

By throwing two dice a large number of times (e.g. 360) find how near you come to the probabilities you found in the table.

If you threw 600 times, how many times should you get a score of

a) 3,   b) 4,   c) 7,   d) 10?

5 List the four possible results of tossing 2 coins simultaneously.

　　a) What is the probability of getting 2 heads?
　　b) What is the probability of getting 1 head and 1 tail?

Working together in a group and throwing two coins together a large number of times, see how close your results come to these probabilities.

6 List the eight possible results of tossing 3 coins simultaneously.
What is the probability of getting 　*a*) 2 heads and 1 tail, 　*b*) 3 heads?

7 A bag contains 5 coloured counters. One counter is picked out and its colour noted. It is then returned to the bag. After 100 draws the results are: blue, 43 times; red, 57 times. How many of each colour do you think there are in the bag?

8 The procedure in question 7 is repeated with a bag containing 10 counters. After 250 draws the results are: white 145, green 26 and pink 79. How many of each colour do you think there are?

9 *a*) The same procedure is carried out again with a bag containing 9 counters. After only 24 draws the results are: green 5, red 8 and yellow 11. How many of each colour do you think there are?
　　b) After 150 draws with the same bag of counters, the results are: green 17, red 46 and yellow 87. How many of each colour do you now think there are?
　　c) Which answer is more likely to be right?

10 From a full pack of playing cards (no jokers) one card is taken out and then returned to the pack. This is repeated.
State whether the following are true or false:

　　a) If you are looking for the ace of spades the chance of taking it out the first time is 1 in 52.
　　b) If you take out a card 52 times and note which card you draw each time, your list should contain the ace of spades once.
　　c) If you are looking for the ace of spades, it should be the 52nd card which you draw.
　　d) If you are looking for the ace of spades, the chance of it being the second card which you draw is 1 in 51.
　　e) If you draw four cards they should be one of each suit.
　　f) If you want any card in the suit of clubs it should be the fourth card you look at.

11 In a packet of sweets 6 are orange flavoured, 8 lemon flavoured, 4 strawberry, 2 lime and 4 blackcurrant. List the probabilities of getting each flavour if you only take one sweet.

　　a) Which is the most likely?
　　b) Which is the least likely?
　　c) Add all the probabilities together. What do you find?

12 A badly worn coin is known to be twice as likely to show heads as tails.

　　a) What is the probability of tossing a head?
　　b) What is the probability of getting a tail?

13 Weather records show that the chance of the temperature falling below freezing point during a certain week in the year is 3/8. On one night during that week what is the probability of there not being any frost?

14 A bag contains 32 coloured beads.

a) The probability of taking out a blue bead is 1/8. How many blue ones are there?
b) There are twice as many pink beads as there are blue. What is the probability of the first one you take out being pink?
c) If the rest are white, what is the probability of the first one being white?

15 A question in an examination paper is divided into two parts and there are four answers given in each part. You have to put a ring round the answer you think is correct in each part. If you don't understand the question and just guess, what is the chance of ringing both the right answers?

16 In question 15, what is the chance of ringing

a) one right answer only,
b) at least one right answer?

(*Hint* Calling the answers *a*, *b*, *c* and *d* in part 1, and *e*, *f*, *g* and *h* in part 2, the correct ones being *a* and *e*, write down every possible combination such as *ae*, *af*, *ag*, etc.)

5B

1 A box contains 5 pieces of white chalk and 4 pieces of coloured.

a) If one piece is taken out, what is the probability that it is white?
b) If it was a white piece and it is not returned to the box, what is the probability that the next piece is also white?

2 A box of 12 pencils contains 3 broken ones.

a) If one is taken out, what is the probability that it is a good one?
b) If it was a broken one and is not returned to the box, what is the probability that that the next one is also broken?

3 A box of chocolates has 5 with hard centres and 7 with soft centres. In another box containing 18 chocolates, 8 have hard centres. If one chocolate is selected at random from each box, in which box is the probability of selecting a soft centre the greater?

4 A book with 234 numbered pages has nine picture pages. What is the probability of opening the book at a picture?

5 In a class of 30 girls, four wear spectacles. A book monitress is chosen at random. What is the probability that the monitress wears spectacles?

6 In a box of 10 different coloured pencils the red one is broken.

a) If I take a pencil out without looking, what is the probability that it is the red one?
b) If I put this one to one side and take another, what is the probability that this second one is the purple one?

7 In a box of 100 drawing pins, 5 are bent. I select one at random.

a) What is the chance of this one being bent?
b) If it is bent and I then take a second without replacing the first, what is the probability that this one is also bent?

8 A class of 28 have each been given a new textbook. It is then found that 4 of the books have pages missing. What is the probability that the last boy on the register has been given a faulty book?

9 To decide which of two players starts a game, a pack of cards is cut.

a) What is the probability of the first player cutting an ace?
b) If he actually cuts at the king of spades, and does not remove the card, what is the probability of the second player cutting an ace?
c) If the first player had withheld the king of spades, would the second one have been more likely to get an ace?

10 A letter is chosen at random from these five: APRIL.

a) What is the probability that it is a vowel?
b) What is the probability that it is the letter P?
c) What is the probability that it is not P?

11 a) In how many different ways can the three letters A, B, C be arranged? List these arrangements.
If the three letters are written down at random,
b) what is the probability that A is in the middle?
c) what is the probability of them being in alphabetical order?

12 In a 'multiple choice' question five possible answers are given. If a candidate decides to guess the answer, what chance has he of being right?

13 In an examination question with three answers given, the candidates have to state in each case whether the answer given is 'true' or 'false'. List all the possible solutions. What is the chance of being right?

14 Mrs Black bought five raffle tickets hoping to win the first prize.

a) If 120 tickets were sold altogether, what chance had Mrs Black of winning?
b) If there were four prizes in the draw, and the total number of tickets was restricted to 120, how many tickets should Mrs Black have bought to be sure of a prize?

15 Out of a group of 28 people, how many would you expect to have been born on a Friday?

16 Out of the set of numbers from 10 to 18 inclusive, one is chosen at random. What is the probability that it is

a) prime   b) a multiple of 4   c) a factor of 72?

17 A letter is selected at random from the capital letters of the alphabet. What is the probability that it is

a) a vowel,
b) a consonant,
c) a letter with a vertical line of symmetry,
d) a letter with a horizontal line of symmetry,
e) a letter with half turn rotational symmetry,
f) a letter without symmetry?

Why is the sum of your answers to parts c, d, e, f not equal to one?

18 A letter is chosen at random from the word SEED.
What is the probability that it is an E?
List all the possible ways of arranging the four letters.
If one arrangement is chosen at random, what is the probability that

    a) it spells seed,
    b) the letters are in alphabetical order,
    c) the two E's come together?

19 Mrs Purchase is in a hurry to pay for her groceries when leaving the supermarket. There are five checkout points altogether, and she counts three people at each of three of them and four at each of the other two. She chooses one with three waiting. What is the chance that she has chosen the swiftest moving queue? Discuss.

20 The results of an examination for 80 pupils were as follows:

Mark	0–10	11–20	21–30	31–40	41–50	51–60	61–70	71–80	81–90	91–100
No. of candidates	2	5	8	9	12	14	20	6	3	1

    a) A distinction is awarded for a mark over 80%. What is the probability of a candidate having a distinction?
    b) What is the probability of having a mark over 60 which is not a distinction?
    c) What is the probability of having a mark of 30 or less?
    d) A mark of 40 or less is classed as grade F. If a boy is given a grade F, what is the probability that his mark is greater than 30?

5C

1 *a*) A drawer contains a pair of white socks and a pair of blue, but the four socks are loose in the drawer and not in pairs. If I take two socks out of the drawer when it is too dark to see the colour, what is the probability that I shall get a matching pair? (You can find this answer by listing all the possible ways of choosing two socks, but you may be able to find a quicker way.)
    b) If I particularly want the blue pair, what is the probability that I shall get them?

2 What is the probability of taking out a matching pair if there are three pairs in the drawer to begin with – one white, one blue and one striped?

3 Imagine dice in the shape of triangular based pyramids so that the numbers go from one to four instead of one to six. If two of these dice, one red and one blue, are thrown together how many possible ways of scoring are there?
What is the probability of *a*) scoring 8, *b*) scoring 3, *c*) both the dice showing the same number, *d*) the total score being an odd number, *e*) the score on the blue one being more than the score on the red one?

4 In a set of six cups, two are cracked.

    a) What is the probability of getting the cracked pair if two are lifted off the shelf without first being checked?
    b) What is the probability of the two being whole?

5 There are five telephones at a bus station. If two are out of order, what is the probability of the first two I try both being out of order?

6 A family consisting of mother, father and two children sit in a row to have their photograph taken. In how many different ways can they be arranged?
If one photograph is taken of each of the possible arrangements and one is picked at random, what is the probability of *a*) the two children being in the middle, *b*) the parents sitting together, *c*) father being on the end?

7 *a*) Four rose bushes are planted in a row. They have all lost their labels but it is known that three are pink and one is yellow. What is the probability that the three pink ones come together in the row?
b) If five rose bushes are planted, three pink and two yellow, what is the probability of the three pink being next to one another?

8 Linda, Gail, John and Simon have consecutive numbers on their seat tickets for a concert. What is the probability that Linda will be sitting next to John?

9 Mr and Mrs Cash and Mr and Mrs Banks sit round a table to play cards. What is the probability of Mr and Mrs Cash sitting next to one another?

10 In a competition in a magazine, five pictures are shown and eight captions are given.

a) What chance have I of guessing the correct caption for the first picture?
b) If I am correct, what is the probability of guessing the correct caption for the second picture?
c) What is the probability of guessing both of these correctly?
d) What is the probability of guessing all five correctly?

6 Combined Transformations

6A Translations

1 Give the co-ordinates of the image of (0,0) under these separate translations:

a) $\begin{pmatrix} 2 \\ 1 \end{pmatrix}$ b) $\begin{pmatrix} -3 \\ 2 \end{pmatrix}$ c) $\begin{pmatrix} -2 \\ -4 \end{pmatrix}$ d) $\begin{pmatrix} 0 \\ -3 \end{pmatrix}$ e) $\begin{pmatrix} 5 \\ -2 \end{pmatrix}$

2 Give the co-ordinates of the image of (1,3) under these separate translations:

a) $\begin{pmatrix} 3 \\ 1 \end{pmatrix}$ b) $\begin{pmatrix} 2 \\ -2 \end{pmatrix}$ c) $\begin{pmatrix} -1 \\ 0 \end{pmatrix}$ d) $\begin{pmatrix} -4 \\ 1 \end{pmatrix}$ e) $\begin{pmatrix} 0 \\ 5 \end{pmatrix}$

3 a) If the translations T_1, T_2, T_3 are defined by the vectors $\begin{pmatrix} 4 \\ 3 \end{pmatrix} \begin{pmatrix} -3 \\ -1 \end{pmatrix} \begin{pmatrix} 1 \\ -2 \end{pmatrix}$

respectively and A is the point $(4,-2)$ find the co-ordinates of $T_1(A)$, $T_2(A)$, $T_3(A)$.
b) Find also the positions of $T_2T_1(A)$, $T_1T_3(A)$, $T_3T_1(A)$ and $T_2T_3(A)$ where $T_2T_1(A)$ means 'Translate A by T_1 and translate $T_1(A)$, the image of A, by T_2'.

4 If B is the point $(-1,2)$ and T_1 is represented by the vector $\begin{pmatrix} 0 \\ -2 \end{pmatrix}$, T_2 by $\begin{pmatrix} -1 \\ 3 \end{pmatrix}$

and T_3 by $\begin{pmatrix} -2 \\ -3 \end{pmatrix}$ find the positions of:

a) $T_1(B)$ b) $T_2T_1(B)$ c) $T_3(B)$ d) $T_1T_3(B)$ e) $T_3T_1(B)$

5 $TT(A)$ can be written $T^2(A)$. Similarly $T^3(A)$ is the same as $TTT(A)$. If T denotes the translation $\begin{pmatrix} 2 \\ 1 \end{pmatrix}$ and A is the point (4,3), find the co-ordinates of $T(A)$, $T^2(A)$, $T^3(A)$.

6 T_1 denotes the translation $\begin{pmatrix} 1 \\ 0 \end{pmatrix}$, T_2 denotes the translation $\begin{pmatrix} 0 \\ -2 \end{pmatrix}$ and T_3 $\begin{pmatrix} -3 \\ 1 \end{pmatrix}$.

If X is the point $(1,-4)$ find the co-ordinates of

a) $T_1{}^2(X)$ b) $T_2T_1(X)$ c) $T_3{}^3(X)$ d) $T_1T_3{}^3(X)$
e) $T_2{}^2(X)$ f) $T_3T_2{}^2(X)$

7 If the direction of the translation T is reversed then the translation is written T^{-1}.

a) Using the translation vectors given in question 6, express as vectors $T_1{}^{-1}$, $T_2{}^{-1}$, $T_3{}^{-1}$.
b) If Y is the point (1,1) find the co-ordinates of $T_1{}^{-1}(Y)$, $T_2{}^{-1}(Y)$, $T_3{}^{-1}(Y)$.
c) Write down the vectors which define the translations $T_1{}^{-2}$, $T_2{}^{-2}$, $T_3{}^{-3}$.

8 If L, M and N are the translations $\begin{pmatrix} -1 \\ 3 \end{pmatrix} \begin{pmatrix} 4 \\ -2 \end{pmatrix} \begin{pmatrix} -2 \\ 1 \end{pmatrix}$
write the single vectors which define these combined translations:

a) LM b) L^2 c) L^2N d) N^{-1} e) N^{-2}
f) $N^{-2}M$ g) M^2 h) NM^2 i) N^2M

9 Given the points $A(1,-4)$, $B(-2,1)$, $C(2,-2)$ and the translations X, Y, Z defined by $\begin{pmatrix} 1 \\ -1 \end{pmatrix} \begin{pmatrix} 3 \\ 2 \end{pmatrix}$ and $\begin{pmatrix} 0 \\ 4 \end{pmatrix}$ respectively, find the positions of:

a) $X(A)$ b) $X^{-1}(A)$ c) $YX^{-1}(C)$ d) $Z^2(B)$ e) $Y^{-2}(B)$
f) $YZ(C)$ g) $X^3(B)$ h) $Z^{-1}(C)$ i) $Y^{-1}Z^{-1}(C)$ j) $Z^3(A)$

10 a) Referring to the translations X, Y, Z of question 9, write down the single vector equivalent to the translations XX^{-1}, YY^{-1}, $Z^{-1}Z$.
b) Is it true to say that $TT^{-1} = T^{-1}T = I$ where I is the identity and T is any translation?
c) The identity translation is the one which leaves the object on which it operates unchanged. Give the vector which is equivalent to I.

11 a) If P, Q, R are translations denoted by $\begin{pmatrix} 4 \\ -1 \end{pmatrix} \begin{pmatrix} -3 \\ 2 \end{pmatrix} \begin{pmatrix} 1 \\ 3 \end{pmatrix}$ respectively, write down as single vectors PQ, QP, PR, RP.
Are translations commutative?
b) Write as single vectors QR, PQR. Is the answer the same if you find PQ then PQR? Are translations associative?

12 If the translations T_1 T_2 T_3 are denoted by the vectors $\begin{pmatrix} 1 \\ 2 \end{pmatrix} \begin{pmatrix} -1 \\ 2 \end{pmatrix} \begin{pmatrix} 3 \\ -2 \end{pmatrix}$ and P is the point $(4,2)$, find:

a) the co-ordinates of A if $T_1T_2(P) = A$
b) the co-ordinates of B if $T_3(B) = P$
c) the co-ordinates of C if $T_3^{-1}(C) = P$
d) the co-ordinates of D if $T_2^2(D) = P$
e) the co-ordinates of E if $T_1^{-2}(P) = E$

13 Plot the three points $A(1,1)$, $B(4,1)$, $C(1,3)$ and join up the triangle ABC. Draw the image of this triangle after translation by the vector $T_1 \begin{pmatrix} 1 \\ -2 \end{pmatrix}$. Translate this second triangle by the vector $T_2 \begin{pmatrix} -3 \\ -1 \end{pmatrix}$. Write down the translation vector T_3 which would transform the first triangle into the third. What can you say about T_1, T_2 and T_3? Write down also the vector T_4 which would translate the third triangle on to the first. What can you say about T_1, T_2 and T_4?

14 $T_3T_2T_1$ can be arranged in six different ways. What are they? As translations are both commutative and associative, these three translations must always give the same result whatever the order in which they are carried out.

Taking T_1 as $\begin{pmatrix} 1 \\ 2 \end{pmatrix}$, T_2 as $\begin{pmatrix} 2 \\ 3 \end{pmatrix}$ and T_3 as $\begin{pmatrix} 3 \\ -1 \end{pmatrix}$ find the co-ordinates of $T_2T_3T_1(P)$ and $T_1T_3T_2(P)$ where P is the point $(2,2)$. Your answer should be the same for both. What is it?

* 15 Operations represented by a string of translations can often be much simplified by rearranging. Thus $T_3{}^{-1}T_2T_3$ can be rearranged as $(T_3{}^{-1}T_3)T_2$ which is IT_2 which is T_2. Similarly $T_2{}^2T_1T_2{}^{-2}T_3$ can be rearranged as $(T_2{}^2T_2{}^{-2})T_1T_3$ which is $I^2T_1T_3$ which is T_1T_3. Simplify

a) $T_4T_3{}^2T_4{}^{-1}$

b) $T_2{}^{-2}T_1T_2T_1{}^{-1}$

c) $T_3{}^2T_2T_1{}^2T_2{}^{-1}T_3{}^{-1}$

d) $T_2T_3T_1T_2{}^{-1}T_3{}^{-1}$

e) $T_1{}^3T_2T_1{}^{-2}T_2T_1{}^{-1}$

* 16 If P is the point $(2,2)$ and T_1, T_2, T_3 are the translations $\begin{pmatrix}3\\1\end{pmatrix}$, $\begin{pmatrix}-2\\1\end{pmatrix}$, $\begin{pmatrix}1\\0\end{pmatrix}$ find the co-ordinates of

a) $T_1T_2{}^{-1}T_1{}^{-1}(P)$

b) $T_2T_1{}^2T_2{}^{-1}(P)$

c) $T_1T_2T_3{}^{-1}(P)$

d) $T_2{}^{-2}T_3T_2(P)$

e) $T_1T_2{}^2T_3T_2{}^{-2}T_1{}^{-1}(P)$

* 17 a) If X, Y, Z and W are the translations $\begin{pmatrix}4\\2\end{pmatrix}$, $\begin{pmatrix}-3\\1\end{pmatrix}$, $\begin{pmatrix}-4\\-2\end{pmatrix}$ and $\begin{pmatrix}3\\-1\end{pmatrix}$

respectively, what can you say about $WZYX$?

b) We have already seen that $WZYX(A)$ is unchanged if the order of the translations is changed, but the path traced out during the translations will be different for different orders. If A is the point $(3,2)$, give the co-ordinates of the vertices of the paths traced out during the translations $WZYX(A)$ and $ZXYW(A)$.

c) Draw rough sketches of the two paths in b.

d) Describe geometrically the paths traced out by the translations $YZWX(A)$ and $YWXZ(A)$.

* 18 a) If Z is $\begin{pmatrix}-2\\-1\end{pmatrix}$, W is $\begin{pmatrix}1\\-2\end{pmatrix}$ and X and Y have the same values as in question 17,

what can you say about $WZYX$?

b) If B is the point $(1,2)$ describe geometrically the path traced out during the operation $WZYX(B)$.

c) Give rough sketches of the paths traced out during the operations $WYZX(B)$ and $ZXWY(B)$.

** 19 There are 24 different orders in which four given translations can be carried out. If $WZYX=I$, how many different shapes will these translations give

a) if two pairs are parallel as in question 17,

b) if only one pair is parallel as in question 18,

c) if no pairs are parallel?

(Shapes that can be transformed into one another by reflection, rotation, translation or a combination of all these are considered as the same shape.)

** 20 'A necessary condition that the path followed during the translations $WZYX$ forms a closed polygon is that $WZYX=I$. Is this a sufficient condition?' This is the mathematician's way of asking a question. It means, 'Whenever the path followed during the translations $WZYX$ is a closed polygon, then $WZYX=I$; but if $WZYX=I$, is the path followed during the translations necessarily a closed polygon?' What do you think?

6B Enlargements

1 Let E_1 be an enlargement with centre $(1,0)$ and scale factor 3. Let E_2 be an enlargement with centre $(0,2)$ and scale factor 2. Let them operate on the unit square L whose vertices are $(1,1)$, $(2,1)$, $(2,2)$, $(1,2)$.

a) On graph paper using a scale of 1 cm to 1 unit draw L and find the co-ordinates of the vertices of $E_1(L)$. Find also the co-ordinates of $E_2E_1(L)$ (where $E_2E_1(L)$ is the image of $E_1(L)$ under the enlargement E_2).
b) Find similarly the co-ordinates of $E_2(L)$ and $E_1E_2(L)$.
c) Is $E_1E_2(L)$ the same as $E_2E_1(L)$, i.e. are E_1 and E_2 commutative?

2 Repeat question 1 but with both E_1 and E_2 having $(0,2)$ as the centre of enlargement and everything else unchanged.
Are E_1 and E_2 commutative this time?

3 If E_1 is an enlargement with centre $(-1,0)$ and scale factor 2, and E_2 is an enlargement with centre $(0,-1)$ and scale factor 3, and L is the unit square used above, find *a)* $E_1(L)$ and $E_2E_1(L)$ *b)* $E_2(L)$ and $E_1E_2(L)$ *c)* Are E_1 and E_2 commutative?

4 Repeat question 3 but this time let both E_1 and E_2 have the centre $(-1,0)$. Are E_1 and E_2 commutative?

* 5 From questions 1 to 4 you have probably gathered that enlargements are commutative if they have the same centre, but they are not commutative if they have different centres. This is a general rule, but there is one (rather trivial) exception. Can you see what it is?

6 *a)* If T is the translation $\begin{pmatrix} 2 \\ 3 \end{pmatrix}$, E the enlargement with centre $(1,0)$ and scale factor 3 and L the unit square $(1,1)$, $(2,1)$, $(2,2)$, $(1,2)$, find the co-ordinates of i) $E(L)$, ii) $TE(L)$, iii) $T(L)$, iv) $ET(L)$.
b) Are $ET(L)$ and $TE(L)$ the same, i.e. are T and E commutative?

7 Repeat question 6 but with E an enlargement of 2 about $(0,2)$ and T the translation $\begin{pmatrix} 3 \\ 4 \end{pmatrix}$.

* 8 Translations and enlargements are not commutative. There are however two (rather trivial) exceptions. Can you see what they are?

6C Combining Rotations

In this exercise all rotations are positive if anti-clockwise. Q_1, H and Q_3 are rotations of $90°$, $180°$ and $270°$ about the origin (i.e. quarter turns, half turns and three quarter turns about the origin). Other rotations are denoted by R, with or without suffixes.

Combining Rotations about the Same Centre

1 On a sheet of graph paper mark the triangle formed by the points $(4,1)$, $(4,0)$ and $(2,0)$. Shade it lightly and call it P.

a) Mark in the positions of $Q_1(P)$, $H(P)$ and $Q_3(P)$.
b) $Q_1{}^2(P)$ is $H(P)$. What are $Q_1{}^3(P)$, $HQ_1(P)$, $H^2(P)$, $Q_1H(P)$, $Q_1Q_3(P)$ and $Q_3Q_1(P)$?
c) What are $Q_1{}^2$, $Q_1{}^3$, HQ_1, H^2, H^3, Q_1H, Q_1Q_3, Q_3Q_1, $Q_3{}^2$?
d) Looking at your answers to *b* and *c*, are combinations of pairs of Q_1, H and Q_3 commutative?
e) Would it be true to say that in a combined rotation of the type $Q_1HQ_3{}^2$ we can perform the operations in any order we like and still get the same result?

40

2 Give another symbol for $Q_1^{-1}, H^{-1}, Q_3^{-2}, H^{-2}, Q_1^{-3}$

3 Simplify

a) $Q_1^2 H$ b) $Q_1^2 H^{-1}$ c) $Q_3^2 H^{-1}$ d) H^3 e) $H^2 Q_1^4 Q_3$

f) Q_1^3 g) Q_3^3 h) Q_1^4 i) H^4 j) Q_3^4

4 P is the point $(3,0)$ and R is a rotation of $60°$ about the origin. Using graph paper and a protractor or set square plot as accurately as you can the positions of a) $R(P)$ b) $R^2(P)$ c) $R^3(P)$. In each case give their co-ordinates.   d) What power of R is equal to I?

5 If P is $(3,0)$ and R is a rotation of $60°$ about $(3,3)$, find by accurate drawing the co-ordinates of a) $R(P)$ b) $R^2(P)$ c) $R^3(P)$.   What power of R is equivalent. to I?

6 If P is $(2,2)$, R_1 a rotation of $45°$ about $(0,3)$ and R_2 a rotation of $60°$ about $(0,3)$, find by accurate drawing   a) $R_1(P)$ b) $R_2(P)$ c) $R_1R_2(P)$ d) $R_2R_1(P)$. Are R_1 and R_2 commutative?

7 If P is the point $(2,1)$, R_1 is a rotation of $40°$ about $(1,1)$ and R_2 a rotation of $50°$ about the same centre $(1,1)$, find by accurate drawing the co-ordinates of   a) $R_1(P)$ b) $R_2(P)$ c) $R_1R_2(P)$.   d) $R_2R_1(P)$.   Are R_1 and R_2 commutative?

8 If R_1, R_2, R_3 are rotations of $30°$, $40°$ and $50°$ respectively about the origin, describe in words

a) $R_1R_2^2R_3^{-2}$ b) $R_2R_3R_1^2$ c) $R_3^{-1}R_2^2R_1^{-2}$

d) $(R_2R_1^{-1})^2$ e) $R_1^{-1}R_2^3R_3^{-2}$ f) $R_1^2R_2^{-2}R_3^{-1}$

Example $R_1^2R_2^2R_3^{-1}$ is a rotation of $90°$ about the origin.

Rotations about Different Centres

9 If A is the unit square $KLMN$ shown in the diagram, R_1 is a rotation of $90°$ about $(1,0)$ and R_2 is a rotation of $-90°$ about $(1,1)$,

a) sketch the positions of $R_1(A)$ and $R_2R_1(A)$, lettering the four vertices.
b) to what simple transformation is R_2R_1 equivalent?
c) sketch the positions of $R_2(A)$ and $R_1R_2(A)$, lettering the four vertices.
d) to what simple transformation is R_1R_2 equivalent?
e) are R_1 and R_2 commutative?

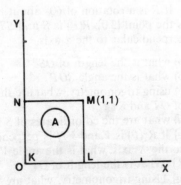

10 Draw sketches to show the position of $R_2R_1(A)$ and $R_1R_2(A)$ for the following values of R_1 and R_2. In each case state the equivalent simple transformation.

a) R_1, half turn about $(2,0)$, and R_2, half turn about $(2,2)$.
b) R_1, three quarter turn about $(0,1)$, and R_2, quarter turn about $(1,0)$.
c) R_1, $90°$ turn about $(1,1)$, and R_2, $-90°$ turn about $(2,0)$.
d) Are R_1 and R_2 commutative?

41

∗11 If R_1 is a quarter turn about the origin, R_2 is a half turn about (2,1) and R_3 is a clockwise three quarter turn about (1,0), transform the arrow shown in the diagram by the transformations stated. In each case state the equivalent single transformation.

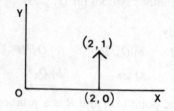

a) $R_3R_2R_1$ b) $R_2R_3R_1$ c) $R_1R_3R_2$

∗12 Can you see why a series of rotations about different points in which the angles of rotation add up to zero or a multiple of 360° is always equivalent to a translation?

∗13 If R_1 and R_2 are rotations of 45° about (2,0) and (3,2) respectively, and P is the point (1,1),

a) find by accurate drawing (or otherwise) the co-ordinates of $R_1(P)$.
b) find the co-ordinates of $R_2R_1(P)$.
c) find the co-ordinates of $R_2(P)$.
d) find the co-ordinates of $R_1R_2(P)$.
e) Are your answers to b and d the same? Are R_1 and R_2 commutative?

∗14 Repeat question 13, letting R_1 and R_2 be rotations of 60° about (3,1) and (4,0) respectively.

∗15 Rotations about the same point are commutative. They can be calculated in any convenient order.
Rotations about different points are not commutative. They must be calculated in the correct order.
Can you think of an exception (a trivial one) to the second statement?

∗ Rotations Using Trigonometry

16 If R is a rotation of 60° about the origin, P is the point (3,0), $R(P)$ is S and ST is perpendicular to the x axis,

a) what is the length of OS?
b) what is the angle SOP?
c) using trigonometry, what are the lengths of OT and ST?
d) what are the co-ordinates of S?
e) If $R^2(P)$ is V and VN is perpendicular to the X axis, what is the angle VON?
f) What is the length of OV?
g) Using trigonometry, what are the lengths of ON and VN?
h) What are the co-ordinates of V?

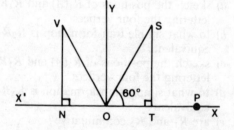

17 If R is a rotation of 50° about O, and Q is the point (4,0), using the method of question 16 find the co-ordinates of a) $R(Q)$, b) $R^2(Q)$ and c) $R^3(Q)$. Check your answer by an accurate drawing.

18 If R is a rotation of $45°$ about S $(2,0)$, Q is the point $(2,2)$ and W is $R(Q)$,

a) what is the length of SW?
b) what are the angles WSQ, WSO?
c) what are the lengths of WN, SN if WN is perpendicular to OX?
d) what are the co-ordinates of W, i.e. of $R(Q)$?
e) calculate the co-ordinates of $R^2(Q)$.
f) calculate the co-ordinates of $R^3(Q)$.

(*Hint* for *e* and *f*, the answer can be obtained without further calculation.)

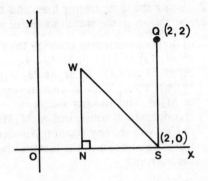

19 If R is a rotation of $-30°$ about S $(0,2)$, Q is the point $(3,2)$ and W is $R(Q)$,

a) what is the length of SW?
b) what is the angle WSO?
c) what are the lengths of WN, SN if WN is perpendicular to the y axis,
d) what are the co-ordinates of W?
e) calculate the co-ordinates of $R^2(Q)$.
f) calculate the co-ordinates of $R^3(Q)$.

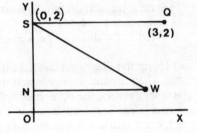

20 If R is a rotation of $70°$ about S $(2,2)$, Q is the point $(4,0)$, W is $R(Q)$, ST is parallel to the x axis and WN is perpendicular to ST,

a) what is the length of SQ?
(Use Pythagoras' Theorem.)
b) what is the length of SW?
c) what is the angle SQO?
d) what is the angle QST?
e) what is the angle WSN?
f) what are the lengths of SN and WN?
g) what are the co-ordinates of W, i.e. of $R(Q)$?

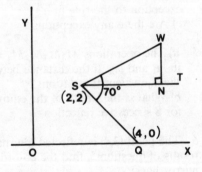

6D Reflections

1 You are given two mirror lines, $x=3$ and $x=6$. Call them m_1 and m_2. Call reflections in these lines M_1 and M_2. You are also given a point P whose co-ordinates are $(1,3)$. Give the co-ordinates of:

a) $M_1(P)$
b) $M_1{}^2(P)$
c) $M_2(P)$
d) $M_2{}^2(P)$
e) $M_2M_1(P)$
f) $M_1M_2M_1(P)$
g) $M_2M_1M_2M_1(P)$
h) $M_2{}^2M_1{}^2(P)$
i) $M_1M_2(P)$
j) $M_2M_1M_2(P)$
k) $M_1M_2M_1M_2(P)$

2 Using the same mirror lines and the arrow shown in the sketch as the object (A),

a) draw one diagram showing the position of all of i) $M_1(A)$ ii) $M_2M_1(A)$
iii) $M_1M_2M_1(A)$ iv) $M_2M_1M_2M_1(A)$
b) State a simple transformation equivalent to M_2M_1 and another simple transformation equivalent to $M_2M_1M_2M_1$.
c) Can you see any relation between these transformations and the distance between the mirrors?

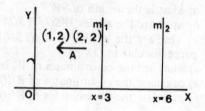

3 a) If m_1 is the line $y=4$, m_2 is the line $y=1$, and P is the point (4,3), give the co-ordinates of $M_2M_1(P)$ and $M_1M_2(P)$.
b) State a translation which is equivalent to M_2M_1 and another translation which is equivalent to M_1M_2.
c) What is the distance between the mirror lines?

* 4 a) Using the diagram shown, is it true to say that when a point P is reflected in two parallel mirrors, the combined reflection is equivalent to a translation in a direction perpendicular to the mirror lines and through a distance which is double the distance between the mirror lines?
b) If the point P is actually on one of the mirror lines, does this constitute an exception to the rule in a?
c) Are there any exceptions?

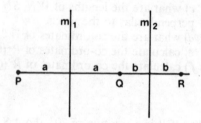

* 5 a) The operations $M_2M_1M_2M_1$ and $M_1M_2M_1M_2$ consist of four successive reflections in m_1 and m_2. If the distance between the mirror lines is 3, what is the length of the equivalent translations?
b) What is the length of the equivalent translation for 6 successive reflections and for 8 successive reflections?

* 6 The point (3,4) is reflected alternately in m_1, m_2, m_1, m_2, etc. By using the results of question 5, find the co-ordinates of the image after reflection in the following mirror lines:

a) $x=2$, $x=7$ (4th reflection)
b) $y=1$, $y=3$ (6th reflection)
c) $x=3$, $x=-1$ (4th reflection)
d) $y=4$, $y=5$ (8th reflection)

You may find it best to work the first two reflections in full to find the direction of the equivalent translation, before applying the results of question 5.

* 7 Repeat question 6a and b, reflecting alternately in m_2, m_1, m_2, m_1, etc.

Reflection in Intersecting Mirrors

Note Questions 8 and 9 are revision questions.

44

8 The point Q is the image of the point P after reflection in the mirror line m. If PQ cuts m in G and O is any point on m,

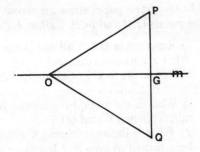

a) what can you say about PG and QG?
b) what are angles OGP and OGQ?
c) what can you say about angles POG and QOG?
d) what can you say about OP and OQ?

9 a) When the point $(3,4)$ is reflected in the line $x=y$, what are the co-ordinates of its image?
b) Repeat with the points $(2,5)$, $(4,6)$, $(3,1)$.
c) If the point (a,b) is reflected in $x=y$, what are the co-ordinates of its image?

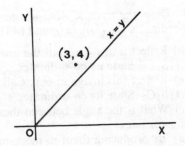

10 The mirrors m_1 and m_2 intersect at O. The image of A in m_1 is B, and the image of B in m_2 is C. AB cuts m_1 at G and BC cuts m_2 at H.

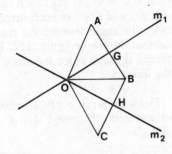

a) What can you say about OA, OB and OC?
b) If the angle AOG is $10°$, what is the angle GOB?
c) If the angle between the mirrors is $30°$, what is angle BOH? What is angle HOC?
d) What is angle AOC? What is the relation between angle AOC and the angle between the mirrors?
e) Would it be true to say in this particular case that the image of A could also be obtained by a rotation about O through twice the angle between the mirrors?

11 The diagram shows a pair of mirror lines at an angle of $60°$ and an arrow AB. Copy the diagram into your book and find CD, the image of AB in the first mirror and EF, the image of CD in the second mirror.
Produce AB and EF until they meet, and measure the angle between them. Is it true to say that EF could have been obtained by a rotation about O of twice the angle between the mirrors?

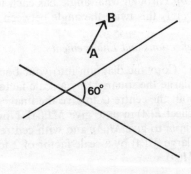

12 Repeat question 11 for the diagram shown.

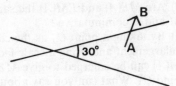

45

13 On graph paper draw an arrow joining the points (2,1) and (4,2). Call it A.

a) Reflect it in OX. Call the image A_1. What are its co-ordinates?
b) Reflect A_1 in OY. Call it A_2. What are its co-ordinates?
c) What is the angle between the two mirror lines OX and OY?
d) What is the angle through which A has been turned to give A_2? Is this angle twice the angle in c?

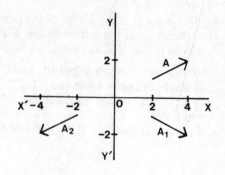

14 Draw the triangle ABC whose co-ordinates are (3,0), (5,0) and (4,1).

a) Reflect it in $y=x$. Call the image $A_1B_1C_1$. State its co-ordinates.
b) Reflect this image in OX. Call it $A_2B_2C_2$. State its co-ordinates.
c) What is the angle between the two mirror lines?
d) By producing them to meet, measure the angle between the lines AB and A_2B_2. Is it twice the angle in c?
e) Repeat d for BC and CA.
f) Is it true to say that each of the lines AB, BC and CA has been turned through twice the angle between the mirror lines, i.e. that the whole figure ABC has been turned through twice the angle between the mirrors?

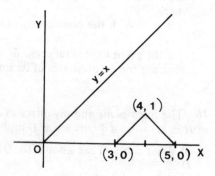

15 Draw the square $ABCD$ whose co-ordinates are (3,2), (4,2), (4,3), (3,3). Reflect it in $y=1$ and letter the image $A_1B_1C_1D_1$. Reflect $A_1B_1C_1D_1$ in $y=x$ and letter the image $A_2B_2C_2D_2$.

a) Give the co-ordinates of $A_2B_2C_2D_2$.
b) Through what angle has each line of $A_2B_2C_2D_2$ been turned?
c) Is this twice the angle between the mirror lines?

Reflections and Enlargements

16 Copy the diagram into your book. Enlarge the square A by a scale factor of 2 about the centre O to give the image $E(A)$. Reflect $E(A)$ in m to give $ME(A)$. Now reflect A in m to give $M(A)$, and with centre O enlarge $M(A)$ by a scale factor of 2 to give $EM(A)$.

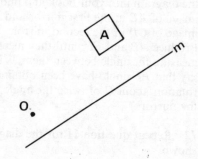

a) Are $ME(A)$ and $EM(A)$ the same? Are M and E commutative?
b) By taking a point O_1 as the centre of enlargement and using a scale factor of 2, $M(A)$ can be enlarged to give $ME(A)$. Find O_1. What can you say about O_1 and O?

17 Make up similar examples for yourself using as centre of enlargement
a) a point *P* which is not on *m*   *b*) a point *Q* which is on *m*.
Are *M* and *E* commutative for *a* or *b*?

18 A rectangle *K* has the co-ordinates (2,1), (4,2), (3, 2·5) and (1, 1·5).

a) If *M* is reflection in *OX* and *E* is enlargement by a scale factor of 3 about (0,1), give the co-ordinates of *EM(K)* and *ME(K)*.
b) Repeat *a* taking the centre of enlargement at the origin.
c) Are *M* and *E* commutative for either *a* or *b*?

19 *a*) Is it true to say that reflections and enlargements are not commutative unless the centre of enlargement lies on the mirror line?
✴ *b*) Could you have deduced this result from your answer to 16 *b*?
✴ *c*) Can you think of two rather trivial exceptions to the rule in *a*?

Reflections and Rotations

20 *R* is a rotation of 60° about a point *O*.
M is a reflection in the mirror line *m*.
T is the rectangle shown.

Copy the diagram into your book and find the positions of *RM(T)* and *MR(T)*. Are *M* and *R* commutative?

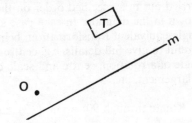

21 Repeat 20 but this time take the centre of rotation at *Q*, a point on the mirror line. Are *M* and *R* commutative?

22 On graph paper draw the triangle *T* with vertices at (2,1), (5,1) and (4,3).

a) If *R* is a rotation of 90° about (0,2) and *M* is reflection in *OX*, give the co-ordinates of the vertices of *MR(T)* and *RM(T)*.
b) Repeat *a* taking the centre of rotation at (0,0).
c) Are *M* and *R* commutative in either *a* or *b*?

23 *a*) Is it true to say that *M* and *R* are not commutative even if the centre of rotation lies on the mirror line? If you are not sure, make up further examples of your own to try and find out.
✴ *b*) Can you think of any exceptions to the rule in *a*?

Reflections and Translations

24 On graph paper, plot the point (5,3). Let *M* be reflection in the *x* axis and *T* the translation $\begin{pmatrix} 4 \\ 2 \end{pmatrix}$.
Find the co-ordinates of *T(P)*, *MT(P)*, *M(P)* and *TM(P)*.
Are *M* and *T* commutative?

25 Repeat 24 with the translation $\begin{pmatrix} 4 \\ 0 \end{pmatrix}$. Are *T* and *M* commutative?

26 Repeat question 24 with the mirror lines and translations stated.

a) $y = x, \begin{pmatrix} 3 \\ 1 \end{pmatrix}$ *b*) $y = x, \begin{pmatrix} 3 \\ 3 \end{pmatrix}$ *c*) $x + y = 6, \begin{pmatrix} 0 \\ 2 \end{pmatrix}$ *d*) $x + y = 6, \begin{pmatrix} 3 \\ -3 \end{pmatrix}$

27 *a*) Is it true to say that translations and reflections are not commutative unless the translation is parallel to the mirror line?

⁕ *b*) Can you think of one (rather trivial) exception to this rule?

Note When the translation is parallel to the mirror line, the transformation is known as a 'glide reflection'.

6E Miscellaneous Transformations

In questions 1 to 6 a series of operations is carried out in a specified order on the arrow shown in the diagram. In each case state a single equivalent transformation, being careful to give full details, e.g. centre and angle of a rotation, centre and scale of an enlargement, etc.

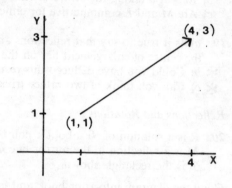

1 Rotate through 90° about (2,3) and then through 180° about (1,5).

2 Rotate through −90° about (0,3) and then through 90° about (3,0).

⁕ *3* Rotate through 90° about (0,0), reflect in $y=x$ and then rotate through −90° about (2,2) and reflect again in $y=x$.

4 Reflect in $x+y=6$ and then reflect in $y=4$.

5 Enlarge by a scale factor of 2 about (2,0) and then by a scale factor of $\frac{1}{2}$ about (0,0).

6 Enlarge by a scale factor of 2 about a centre (1,0), then rotate through 90° about the origin and finally enlarge by a scale factor of 0·5 about a centre (2,1).

7 *ABC* is an equilateral triangle. Its vertices occupy the positions 1, 2, 3 in the plane. R_1 and R_2 are rotations of 120° and 240° about the centre *O*. M_1, M_2 and M_3 are reflections in the lines of symmetry through points 1, 2 and 3.

These five operations, together with the identity *I*, all map the triangle on to itself, but with its vertices in different order. They are known as the 'symmetry operations of an equilateral triangle'.

Copy and complete the table, which shows the positions *A*, *B* and *C* occupy under the six symmetry operations.

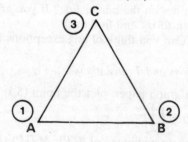

	1	2	3
I			
R_1			
R_2			
M_1	A	C	B
M_2			
M_3			

8 Copy and complete the following table which shows the results of any one of these six operations followed by any other.

1st operation

1 / 2	I	R_1	R_2	M_1	M_2	M_3
I	I		R_2			M_3
R_1						
R_2						
M_1	M_1					
M_2						
M_3						

followed by (rows I, R_1, R_2, M_1, M_2, M_3)

9 *a)* What are the symmetry operations for the rectangle shown in the figure (i.e. what are the operations that map the rectangle on to itself)?
b) Draw up a table similar to the one in question 7 showing the points in space (1,2,3,4) on to which A,B,C,D are mapped by the various operations.
c) Draw up a table similar to the one in question 8 showing the effect of one of these symmetry operations followed by another.

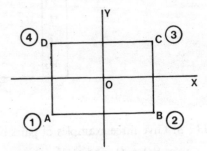

Note Ask your teacher to choose the order in which the columns and rows of the tables in *b* and *c* are to be arranged. This makes checking easier.

10 A square $ABCD$ can be mapped on to itself in eight different ways. A can fall on A, B, C or D. The three other points can follow round from A either clockwise or anticlockwise, giving $4 \times 2 = 8$ arrangements altogether. The eight operations which give rise to the above arrangements are known as the 'symmetry operations of a square'. List them. They include the identity.

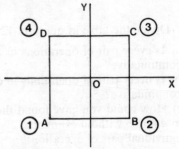

11 Calling the positions occupied by A,B,C,D in the plane 1,2,3,4, copy and complete this table which shows the positions occupied by the square under the eight transformations, the origin being taken at the centre of the square and the axes parallel to the sides. M_1 and M_2 are reflections in $y=x$ and $y+x=0$.

	1	2	3	4
I	A	B	C	D
Q_1				C
H			A	
Q_3		C		
M_x	D			
M_1		D		
M_y			D	
M_2				D

12 When any one of the eight operations is followed by another of the eight, the combined effect is equivalent to yet another of the eight. Copy and complete the following table which shows the effect of combining these operations in pairs. The table in question 11 will help you considerably.

1st operation

	I	Q_1	H	Q_3	M_x	M_1	M_y	M_2
I	I	Q_1	H	Q_3	M_x			
Q_1	Q_1							
H	H		I					
Q_3					M_2	M_x		
M_x				M_1	I	Q_3		
M_1	M_1							
M_y								
M_2								

followed by (left of rows)

13 *a*) Give three examples of pairs of operations that are commutative,

e.g. $HQ_1 = Q_3$ and $Q_1 H = Q_3$.

b) Give three examples of pairs of operations that are not commutative,

e.g. $M_x Q_3 = M_1$ but $Q_3 M_x = M_2$.

14 Divide the table in question 12 into four equal quadrants.

a) Is every pair of operations in quadrant 1 commutative?
b) Is every pair of operations in quadrant 4 commutative?
c) How could you have found the answer to *a* and *b* without examining every individual pair of operations?

1	2
3	4

15 Every set of operations in the table in question 12 is associative. Give three examples, setting them out as follows:

$$(M_x Q_1)H = M_2 H = M_1 \text{ and } M_x(Q_1 H) = M_x Q_3 = M_1$$

so $(M_x Q_1)H = M_x(Q_1 H)$

*16 *a*) How many symmetry operations are there for a regular pentagon?
b) How many symmetry operations are there for a regular hexagon?
c) How many symmetry operations are there for a regular octagon?

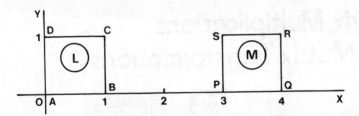

*17 L and M are unit squares. L can be mapped on to M in eight different ways. A can fall on P,Q,R or S, and for each of these B,C and D can follow round the square from A in a clockwise or an anticlockwise direction. This gives the eight ways.

Five of the eight transformations that produce the above mappings are 'single' transformations, and the other three are glide reflections (which can also be thought of as 'single' transformations).

Copy and complete the table below, giving the details of each transformation and the points on to which A,B,C and D are mapped. Under 'details' give vectors, centres and angles of rotation, and equations of mirror lines.

Transformation	Details	A	B	C	D
Translation					
Quarter turn	Centre (2,2), Angle 90°	Q	R	S	P
Half turn					
Three quarter turn					
Reflection					
1st glide reflection	$x+y=2.5$, $\begin{pmatrix} 1.5 \\ -1.5 \end{pmatrix}$	R	Q	P	S
2nd glide reflection					
3rd glide reflection					

*18 Make a similar table for the mapping of unit square F on to unit square J.

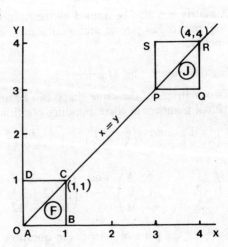

7 Matrix Multiplications and Matrix Transformations

7A Matrix Multiplications

1 A matrix is an array of numbers. It can be square, as shown here $\begin{pmatrix} 2 & 3 \\ 11 & 4 \end{pmatrix}$, rectangular, a single row or a single column.
Describe the following matrices:

a) $\begin{pmatrix} 3 & 1 & 4 \\ 2 & 1 & 6 \\ 1 & 1 & 0 \end{pmatrix}$ b) $\begin{pmatrix} 7 & 2 \\ 3 & 5 \\ 1 & 4 \end{pmatrix}$ c) $(1 \quad 4 \quad 3 \quad 2)$

d) $\begin{pmatrix} 2 \\ 3 \end{pmatrix}$ e) $\begin{pmatrix} 1 & 4 & 4 & 5 \\ 2 & 3 & 1 & 0 \\ 0 & 2 & 4 & 6 \\ 1 & 0 & 1 & 1 \end{pmatrix}$ f) $\begin{pmatrix} 4 & 1 & 6 & 3 \\ 2 & 7 & 1 & 1 \end{pmatrix}$

Can you suggest alternative descriptions for c and d?

2 A matrix can also be named by the number of rows and columns it contains. Thus a $(p \times q)$ matrix has p rows and q columns. Name the six matrices in question 1 using this method.

Adding and Subtracting Matrices

3 Two matrices of the same shape can be added or subtracted. Add or subtract the individual numbers in corresponding positions. Add the following matrices.

a) $\begin{pmatrix} 1 & 1 \\ 2 & 3 \end{pmatrix}$ and $\begin{pmatrix} 3 & 1 \\ 2 & -2 \end{pmatrix}$ b) $\begin{pmatrix} 5 & -3 & 4 \\ 1 & 6 & 7 \end{pmatrix}$ and $\begin{pmatrix} 2 & 2 & 1 \\ 1 & 0 & -3 \end{pmatrix}$

c) $\begin{pmatrix} 1 & 3 & 5 & 4 \\ -2 & 4 & 6 & 0 \\ 5 & 1 & -3 & 0 \end{pmatrix}$ and $\begin{pmatrix} 3 & 5 & -7 & -10 \\ 2 & 2 & 1 & 1 \\ 0 & 1 & 2 & 3 \end{pmatrix}$

4 Subtract the pairs of matrices in question 3. (Take the 2nd of each pair from the first.)

5 Consider the matrix product shown below. Call the 3 matrices *A*, *B* and *C*:

$$\begin{pmatrix} 3 & 4 & 5 \\ 1 & 2 & 3 \\ 1 & 1 & 2 \end{pmatrix} \begin{pmatrix} 2 & 1 & 1 \\ 3 & 1 & 2 \\ 1 & 2 & 1 \end{pmatrix} = \begin{pmatrix} 23 & 17 & 16 \\ 11 & 9 & 8 \\ 7 & 6 & 5 \end{pmatrix}$$
$$(A) \qquad\qquad (B) \qquad\qquad (C)$$

a) Multiply the first row of *A* and the first column of *B* using ordinary 'row and column multiplication': $(3 \quad 4 \quad 5) \begin{pmatrix} 2 \\ 3 \\ 1 \end{pmatrix} = 23$

In what row and column of *C* does 23 appear?

b) Multiply the first row of *A* and the second column of *B* using ordinary 'row and column multiplication'. What is the result? In what row and column of *C* does this result appear?

c) Multiply the second row of *A* and the third column of *B*. What is the result? In what row and column of *C* does this result appear?

d) Would it be true to say that the number in the third row and the second column of *C* is obtained by multiplying the third row of *A* and the second column of *B*?

e) If *m* and *n* are whole numbers not greater than 3, would it be true to say that the number in the *m*th row and the *n*th column of *C* is obtained by multiplying the *m*th row of *A* and the *n*th column of *B*? Give two more examples to illustrate your answer.

6 In question 5, *C* is said to be the matrix product of *A* and *B*. Calculate the following matrix products:

a) $\begin{pmatrix} 2 & 2 \\ 3 & 1 \end{pmatrix} \begin{pmatrix} 4 & 1 \\ 2 & 3 \end{pmatrix}$ b) $\begin{pmatrix} 3 & 2 \\ 4 & 1 \end{pmatrix} \begin{pmatrix} 1 & 1 \\ 4 & 1 \end{pmatrix}$ c) $\begin{pmatrix} 2 & 2 & 3 \\ 1 & 4 & 3 \end{pmatrix} \begin{pmatrix} 4 & 1 \\ 2 & 3 \\ 2 & 1 \end{pmatrix}$

d) $\begin{pmatrix} 8 & 3 & 1 \\ 3 & 2 & 2 \\ 1 & 4 & 1 \end{pmatrix} \begin{pmatrix} 3 & 2 \\ 1 & 2 \\ 1 & 2 \end{pmatrix}$ e) $(1 \quad 1 \quad 2) \begin{pmatrix} 2 & 2 \\ 3 & 0 \\ 0 & 1 \end{pmatrix}$

7 Here are the two matrices we used in question 5 written in reverse order:

$$\begin{pmatrix} 2 & 1 & 1 \\ 3 & 1 & 2 \\ 1 & 2 & 1 \end{pmatrix} \begin{pmatrix} 3 & 4 & 5 \\ 1 & 2 & 3 \\ 1 & 1 & 2 \end{pmatrix}$$

a) Find their product   b) Is the multiplication of these matrices commutative?

8 Reverse the order of the pairs of matrices in question 6 *a* to *e* and find their products. Are any of them the same as the products you found in question 6? Is the multiplication of these pairs of matrices commutative? What happens when you try to multiply *d* and *e* in the reverse order?

9 a) How many terms are there in the first row of a (4 × 3) matrix? Give an example.
b) How many terms are there in the first column of a (3 × 2) matrix? Give an example.
c) The two matrices can be multiplied (1st × 2nd). Why?
d) How many terms are there in each row of a (4 × 2) matrix?
e) How many terms are there in each column of a (3 × 2) matrix?
f) These two matrices cannot be multiplied (1st × 2nd). Why not?

g) Can the following matrices be multiplied? If so, give the products:

i) $\begin{pmatrix} 3 & 2 \\ 1 & 4 \end{pmatrix}$ $\begin{pmatrix} 3 & 1 \\ 3 & 2 \\ 3 & 3 \end{pmatrix}$ ii) $\begin{pmatrix} 1 & 2 & 3 \\ 2 & 3 & 4 \end{pmatrix}$ $\begin{pmatrix} 1 & 2 \\ 1 & 4 \end{pmatrix}$

iii) $\begin{pmatrix} 1 & 0 \\ 0 & 1 \end{pmatrix}$ $\begin{pmatrix} 1 & 1 \\ 0 & 0 \end{pmatrix}$ iv) $(1 \quad 3 \quad 4)$ $\begin{pmatrix} 2 & 3 \\ 3 & 2 \\ 0 & 1 \end{pmatrix}$ v) $\begin{pmatrix} 3 \\ 2 \\ 1 \end{pmatrix}$ $(1 \quad 2 \quad 3)$

h) Can the following matrices be multiplied (in the order stated)?

i) (3×4) matrix $\times (2 \times 1)$ matrix
ii) (5×2) matrix $\times (2 \times 4)$ matrix
iii) (3×3) matrix $\times (3 \times 4)$ matrix
iv) (3×3) matrix $\times (2 \times 2)$ matrix
v) $(p \times q)$ matrix $\times (q \times r)$ matrix

i) What is the condition that a $(p \times q)$ matrix can pre-multiply an $(r \times s)$ matrix?

10 Calculate the following matrix products. Where it is not possible, say so.

a) $\begin{pmatrix} 3 & 1 \\ 0 & 2 \end{pmatrix}$ $\begin{pmatrix} 1 & 1 \\ 0 & 1 \end{pmatrix}$ b) $\begin{pmatrix} 3 & 2 \\ 1 & 2 \end{pmatrix}$ $\begin{pmatrix} 1 & 0 \\ 0 & 1 \end{pmatrix}$ c) $\begin{pmatrix} 4 & 3 \\ 2 & 1 \end{pmatrix}$ $\begin{pmatrix} 1 & 3 & 1 \\ 2 & 4 & 1 \end{pmatrix}$

d) $\begin{pmatrix} 3 & 1 & 1 \\ 3 & 2 & 2 \\ 1 & 3 & 4 \end{pmatrix}$ $\begin{pmatrix} 2 & 3 \\ 1 & 1 \\ 1 & 1 \end{pmatrix}$ e) $\begin{pmatrix} 4 & 1 & 4 & 1 \\ 2 & 0 & 0 & 1 \end{pmatrix}$ $\begin{pmatrix} 1 & 0 & 4 & 4 \\ 2 & 0 & 3 & 3 \\ 1 & 1 & 1 & 1 \\ 2 & 1 & 0 & 1 \end{pmatrix}$

f) $\begin{pmatrix} 4 & 1 & 4 \\ 2 & 0 & 0 \\ 1 & 0 & 0 \\ 1 & 0 & 1 \end{pmatrix}$ $\begin{pmatrix} 1 & 0 & 4 & 4 \\ 2 & 0 & 3 & 3 \end{pmatrix}$

11 Write the matrices in question 10 in reverse order and calculate their products. Where this is not possible, say so.

12 In reversing the order of a matrix product, sometimes the product becomes impossible. Give an example from question 10.

13 You are given the following matrices:

$A = \begin{pmatrix} 1 & 2 \\ 1 & 3 \end{pmatrix}$ $B = \begin{pmatrix} -2 & 3 \\ 1 & -4 \end{pmatrix}$ $C = \begin{pmatrix} 0 & -2 \\ 3 & -1 \end{pmatrix}$ $D = \begin{pmatrix} -1 & 2 & 5 \\ 4 & 0 & -2 \end{pmatrix}$

$E = \begin{pmatrix} 2 & 4 \\ 1 & -3 \\ 2 & 0 \end{pmatrix}$ $F = \begin{pmatrix} 1 & 4 & 3 \\ 2 & 0 & 1 \\ -2 & -1 & 4 \end{pmatrix}$ $G = \begin{pmatrix} 1 & 0 & 1 \\ 0 & -1 & 1 \\ 1 & -1 & 0 \end{pmatrix}$ $H = \begin{pmatrix} 1 & 2 \\ 2 & 1 \end{pmatrix}$

$K = \begin{pmatrix} -2 & 3 \\ 3 & -2 \end{pmatrix}$

Calculate the following matrix products. Where this is impossible, say so.

a) A^2 b) B^2 c) DE d) ED e) EF f) FE g) HK
h) KH i) DG j) DF k) E^2 l) CD m) EC n) AB

14 Is matrix multiplication associative, i.e. does $A(BC)$ equal $(AB)C$ where A, B and C are matrices?

Investigate this for yourself, using the products ABC, CDE and BCH from question 13.

15 Multiply the following pairs of matrices:

a) $\begin{pmatrix} 1 & 0 \\ 0 & 1 \end{pmatrix} \begin{pmatrix} 3 & 1 \\ 2 & 2 \end{pmatrix}$ b) $\begin{pmatrix} 3 & 1 \\ 2 & 2 \end{pmatrix} \begin{pmatrix} 1 & 0 \\ 0 & 1 \end{pmatrix}$

c) $\begin{pmatrix} 1 & 0 & 0 & 0 \\ 0 & 1 & 0 & 0 \\ 0 & 0 & 1 & 0 \\ 0 & 0 & 0 & 1 \end{pmatrix} \begin{pmatrix} 4 & 1 & 6 & 3 \\ 2 & 2 & 1 & 1 \\ 1 & 4 & 1 & 1 \\ 1 & 0 & 0 & 1 \end{pmatrix}$ d) $\begin{pmatrix} 4 & 1 & 6 & 3 \\ 2 & 2 & 1 & 1 \\ 1 & 4 & 1 & 1 \\ 1 & 0 & 0 & 1 \end{pmatrix} \begin{pmatrix} 1 & 0 & 0 & 0 \\ 0 & 1 & 0 & 0 \\ 0 & 0 & 1 & 0 \\ 0 & 0 & 0 & 1 \end{pmatrix}$

16 The matrices $\begin{pmatrix} 1 & 0 \\ 0 & 1 \end{pmatrix}$, $\begin{pmatrix} 1 & 0 & 0 \\ 0 & 1 & 0 \\ 0 & 0 & 1 \end{pmatrix}$ and $\begin{pmatrix} 1 & 0 & 0 & 0 \\ 0 & 1 & 0 & 0 \\ 0 & 0 & 1 & 0 \\ 0 & 0 & 0 & 1 \end{pmatrix}$ etc.

are called unit matrices and denoted by I.

    a) Write in full the (5×5) unit matrix.
    b) Is it true to say that $IA = AI = A$ for any square matrix A?
    c) What number in ordinary arithmetic or algebra corresponds to I in matrix algebra?
    d) I is also called the identity matrix. Why is this?

17 If you multiply a number a in ordinary arithmetic or algebra by 0, then $a \times 0 = 0 \times a = 0$, whatever the value of a.

Can you think of a matrix O such that a similar result holds in matrix algebra, i.e. $O \times A = A \times O = O$ where A is any square matrix? O is called the Null matrix.

* 18   a) Is multiplication of a square matrix by I or O always commutative?
    b) Can you think of another matrix J such that multiplication of a square matrix by J is always commutative?
    c) If the multiplication of two matrices A and B is commutative, what can you say about the shape of A and of B?
    d) Can you find two matrices A and B such that the multiplication of A and B is commutative, but neither of them is I, O or a multiple of I?

* 19 We have seen that if A is a square matrix, $IA = AI = A$. To find out whether a similar result was true for a rectangular matrix it would be necessary to use a different form of I for each of the two multiplications.

$$\text{Let } A = \begin{pmatrix} 1 & 2 & 3 \\ 4 & 5 & 6 \end{pmatrix}$$

    a) Find the value of IA. What form of I must you use?
    b) Find the value of AI. What form of I must you use?
    c) Can you say that $IA = AI = A$?
    d) Is a similar result true for any rectangular matrix?

* 20   a) If A is a rectangular matrix, is it true to say that $OA = AO = O$?
    b) Would you need to use different forms of O for the two multiplications?

7B Using Matrix Multiplication in Trade and Industry

The examples in this section are very much simplified. 'Real life' transactions would generally be much longer and more complicated and would be worked on a computer.

1 Three firms place regular monthly orders with a manufacturer of cleaning materials. The first table gives details of the orders and the second gives the prices.

	LP	LD	SP		£
1st firm	7	12	5	LP	3.80
2nd firm	12	5	1	LD	2.20
3rd firm	2	16	12	SP	1.80

> Key LP Liquid polish in 5 litre containers
> LD Liquid detergent in 5 litre containers
> SP Scouring powder (dozens)

a) Set up a matrix multiplication with the column vector second and find the cost of the monthly order for each firm.
b) State clearly the meaning of each term in your answer.

2 Repeat question 1, re-writing the column vector as a row vector and pre-multiplying by this row vector. The 3×3 matrix must also be re-written with rows as columns and columns as rows. Explain clearly the meaning of each term in your answer.

3 In three successive years the prices of the products quoted in question 1 were as shown in the table below:

	1st year £	2nd year £	3rd year £
LP	3.00	3.50	3.80
LD	1.80	2.05	2.20
SP	1.40	1.60	1.80

a) Set up a matrix multiplication using the firms' orders from question 1 as the first matrix and the prices for the three years as the second matrix. Complete the multiplication; this will give you the cost of the monthly order for each firm for each of the three years.
b) In what row and column of the product matrix would you find the cost to the 2nd firm of its monthly order in the 3rd year?
c) What is the meaning of the figure in the third row and the first column of the product matrix?
d) Label the rows and columns of the product matrix to show their meaning.

4 A coal merchant delivers coal in 50 kg bags. The first table gives the details of deliveries to three customers, all of whom have the option of paying cash on delivery, within one week of delivery or within two months. The second table gives the relevant prices.

	ASN bags	GTC bags	AB bags		COD £	W £	TM £
Mr McBirnie	60	10	60	ASN	3.20	3.30	3.45
Mrs Concord	100	50	0	GTC	2.15	2.25	2.30
Miss Gwillym	0	10	120	AB	3.05	3.15	3.25

Key	ASN Anthracite stove nuts	COD Cash on delivery
	GTC Grade 2 house coal	W Payment within a week
	AB Anthracite beans	TM Payment within two months

56

a) Write down two matrices (taking the first matrix from the first table) and multiply them together to find how much each customer must pay in each of three methods of payment.

b) In what row and column of the product matrix would you find the amount Miss Gwillym must pay if she pays within a week?

c) What is the meaning of the figure in the 2nd row and last column of the product matrix?

5   A dog food manufacturer sells three kinds of dog biscuits, Canine Cookies (*CC*) Dog's Delight (*DD*) and Wagtail Wonders (*WW*). The first table shows the prices of 50 kg cases of these biscuits, packed in either 1 kg bags, 5 kg packets or loose in the case. Three pet shops are considering placing regularly monthly orders which are shown in the second table.

	1 kg £	5 kg £	loose £		CC cases	DD cases	WW cases
CC	8.50	8.00	7.50	Pet shop A	1	3	4
DD	8.25	7.80	7.35	Pet shop B	4	2	1
WW	8.00	7.60	7.20	Pet shop C	6	1	0

Set up a matrix multiplication, using the matrix arising from the second table first, and find what this monthly order would cost each shop packed in the three different ways available.

Label your product matrix clearly so that its meaning can be seen easily.

✳6   The three shops decide to take their monthly orders in more than one kind of pack. The table below shows their final choice.

	CC			DD			WW		
	1 kg	5 kg	loose	1 kg	5 kg	loose	1 kg	5 kg	loose
1st pet shop	1	0	0	1	1	1	1	2	1
2nd pet shop	2	1	1	1	0	1	0	0	1
3rd pet shop	3	2	1	1	0	0	0	0	0

a) To set up a matrix multiplication to give the cost of the monthly order for each shop, you will have to modify the 'price matrix'. The 'order matrix' is 3×9. What shape will the 'price matrix' be? What shape will the product matrix be?

b) Set up the matrices and complete the multiplication.

c) What is the cost of each of the three orders?

7   A publisher produces three paperback books for children: 'Alison in Orbit' (*AO*), 'The Brave Basenji' (*BB*) and 'Cheerful Charlie' (*CC*). All three are printed versions of popular television programmes.

The first table shows the numbers ordered by a village store (*V*), a small town bookshop (*B*) and a supermarket in a large town (*S*). The second table shows the price per copy for single copies, packs of 10 or packs of 100.

	Number ordered				Cost in pence		
	AO	BB	CC		1's	10's	100's
V	1	3	5	AO	65	60	55
B	20	40	50	BB	70	63	57
S	400	200	300	CC	75	69	63

a) Set up a suitable matrix multiplication to find the total cost of each of the three orders.

b) In your product matrix, three terms should be knocked out as they have no meaning. Which are they? (Answer by row and column.) Three other terms have very little meaning, i.e. the figures they state would not be used. Which are they?

c) State the cost of each of the three orders.

8 A toy shop orders the following goods from a wholesaler: 20 chess sets (*C*), 4 talking dolls (*D*) and 300 model cars (*M*). The table gives the costs, which vary according to whether the order is in ones, tens or hundreds.

	Cost per item (£)		
	1's	10's	100's
C	3.00	2.80	2.50
D	11.00	10.50	10.00
M	0.85	0.80	0.70

a) Set up a matrix product to show the cost of the order.
(*Hint* leave out all unnecessary terms.)
✱ *b*) If the manager of a chain of toy shops ordered 243 of *C*, 112 of *D* and 2361 of *M*, set up a matrix product to show the total cost.
Do not calculate the actual product.

✱ **9** Semi-skilled fitters in an engineering works are paid a basic rate of £1.50 an hour for the first 40 hours, overtime at the rate of £2.50 an hour for the next 8 hours and then £3.00 an hour for overtime in excess of 8 hours. They also receive a punctuality bonus of £2 if they do not lose any time at all during the week.
The table shows the total hours worked by each of three fitters. The letter *B* is added if they have qualified for the bonus. Write down a single matrix multiplication which will give the total earnings of each of the three men for that week.

1st man	2nd man	3rd man
46 hours	51 hours *B*	50 hours *B*

How much does each receive?

✱ **10** A plant assembling machine components uses three waggons to transport its finished assemblies to the parent factory. Waggon *A* will carry up to 5 tonnes, *B* up to 10 tonnes and *C* up to 15 tonnes. There are nine types of assembly, whose individual weights are as follows: 48 kg, 72 kg, 140 kg, 200 kg, 221 kg, 264 kg, 301 kg, 445 kg and 602 kg. Waggon *A* takes the three lightest components, Waggon *C* the three heaviest components and Waggon *B* the others. On a certain day the number to be transported on the first run of the three waggons is as follows, the numbers being in the same order as the weights: 24, 20, 6, 15, 13, 12, 5, 15 and 12.

a) Write down a single matrix multiplication which will give you the load on each waggon. (*Hint* the first matrix will be a 3 × 9 with a lot of zeros; the second will be a 9 × 1.)
b) Are any of the waggons overloaded?

Note There is little advantage in using matrix methods for the type of problem in Section 7B unless these problems recur regularly with different numbers each time. But if they do recur frequently, and a computer is programmed, the answer for each set of numbers can be obtained immediately. This is particularly valuable for long strings of entries which are continually changing.

7C Matrix Transformations

1 The triangle shown in the figure is to be transformed by the matrix $\begin{pmatrix} -1 & 0 \\ 0 & 1 \end{pmatrix}$.

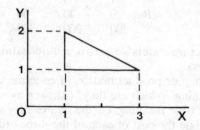

Write the position vectors of the vertices of the triangle as a (2×3) matrix and pre-multiply by the 'operation' matrix.

$$\begin{pmatrix} -1 & 0 \\ 0 & 1 \end{pmatrix} \begin{pmatrix} 1 & 3 & 1 \\ 1 & 1 & 2 \end{pmatrix} = \begin{pmatrix} -1 & -3 & -1 \\ 1 & 1 & 2 \end{pmatrix}$$

a) What are the co-ordinates of the image?
b) Make a rough sketch of both object and image and hence deduce what operation the matrix $\begin{pmatrix} -1 & 0 \\ 0 & 1 \end{pmatrix}$ represents.

2 Operate on the same triangle as in question 1 with the matrices listed below. In each case give the co-ordinates of the image, make a rough sketch showing both object and image, and state what operation the matrix represents.

a) $\begin{pmatrix} 1 & 0 \\ 0 & -1 \end{pmatrix}$ b) $\begin{pmatrix} -1 & 0 \\ 0 & -1 \end{pmatrix}$ c) $\begin{pmatrix} 1 & 0 \\ 0 & 1 \end{pmatrix}$ d) $\begin{pmatrix} 0 & 1 \\ 1 & 0 \end{pmatrix}$

e) $\begin{pmatrix} 0 & -1 \\ 1 & 0 \end{pmatrix}$ f) $\begin{pmatrix} 0 & -1 \\ -1 & 0 \end{pmatrix}$ g) $\begin{pmatrix} 0 & 1 \\ -1 & 0 \end{pmatrix}$

3 Repeat question 2 using as object the arrow whose co-ordinates are (3,1), (3,5), (2,4), (4,4).

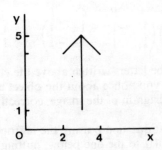

4 You are given the co-ordinates of the vertices of a number of figures. Operate on them with the matrices stated. In each case give the co-ordinates of the image, a rough sketch showing the object and image, and the geometrical operation that the matrix represents.

	Co-ordinates of vertices	Matrix operator
a	(1,1), (3,1), (1,5)	$\begin{pmatrix} 0 & -1 \\ -1 & 0 \end{pmatrix}$
b	(1,1), (4,2) Draw a small arrow head at (4,2)	$\begin{pmatrix} 0 & 1 \\ -1 & 0 \end{pmatrix}$
c	(2,2), (6,2), (6,4), (4,6), (2,4)	$\begin{pmatrix} -1 & 0 \\ 0 & -1 \end{pmatrix}$
d	(1,3), (1,5), (2,4), (4,2), (4,6)	$\begin{pmatrix} -1 & 0 \\ 0 & 1 \end{pmatrix}$

5 Not every (2×2) matrix represents a single simple geometrical transformation. Operate on the rectangle whose co-ordinates are (1,1), (2,1), (2,4) and (1,4) with the matrices stated. In each case give a rough sketch of both object and image and list its co-ordinates.

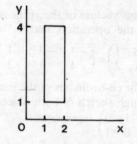

a) $\begin{pmatrix} 1 & 1 \\ 0 & 2 \end{pmatrix}$ b) $\begin{pmatrix} 2 & 1 \\ 0 & 2 \end{pmatrix}$ c) $\begin{pmatrix} 1 & 0 \\ 1 & 2 \end{pmatrix}$

You will be able to see that none of these gives a simple transformation, but a combination of several transformations (enlargements, shears, etc.). See also question 19.

6 Returning to the 2×2 matrices which *do* give a single geometrical transformation, operate on the square $ABCD$ whose co-ordinates are $A(1,1)$, $B(-1,1)$, $C(-1, -1)$, $D(1,-1)$ with the matrix $\begin{pmatrix} -1 & 0 \\ 0 & 1 \end{pmatrix}$.

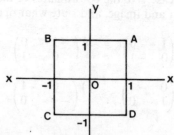

The actual multiplication is

$$\begin{pmatrix} -1 & 0 \\ 0 & 1 \end{pmatrix} \overset{\substack{A \quad B \quad C \quad D}}{\begin{pmatrix} 1 & -1 & -1 & 1 \\ 1 & 1 & -1 & -1 \end{pmatrix}} = \overset{\substack{A \quad B \quad C \quad D}}{\begin{pmatrix} -1 & 1 & 1 & -1 \\ 1 & 1 & -1 & -1 \end{pmatrix}}$$

(Notice the letters written above the columns of the object and image matrices.)
What do you notice about the object and the image?
Draw a diagram of the image, correctly lettered.

7 Repeat question 6 using the matrices listed in question 2. Enter your results in a table similar to the one below, putting the results from question 6 in the first line.

Matrix	Product Matrix	Sketch of Image (correctly lettered)	Geometrical Transformation
$\begin{pmatrix} -1 & 0 \\ 0 & 1 \end{pmatrix}$	$\overset{\substack{A \quad B \quad C \quad D}}{\begin{pmatrix} -1 & 1 & 1 & -1 \\ 1 & 1 & -1 & -1 \end{pmatrix}}$	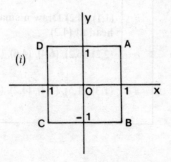	Reflection in y axis
$\begin{pmatrix} 1 & 0 \\ 0 & -1 \end{pmatrix}$			

8 If the object in question 6 has been transformed to give image (*i*),
  a) what transformation was used?
  b) From the table in question 7, what matrix was used?

c) Repeat for image (ii).
d) Repeat for image (iii).

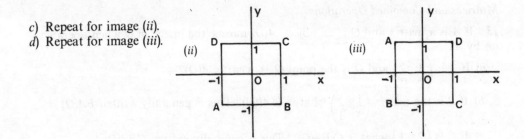

9 (3,2) is the vertex A of an octagon $ABCDEFGJ$ (lettered anti-clockwise) which has the x axis, the y axis, $x=y$ and $y=-x$ as axes of symmetry.

a) List the co-ordinates of the remaining vertices and draw a rough sketch of the octagon.

b) Operate on the octagon with the following matrices:

$$\text{(i)} \begin{pmatrix} 1 & 0 \\ 0 & -1 \end{pmatrix} \qquad \text{(ii)} \begin{pmatrix} -1 & 0 \\ 0 & -1 \end{pmatrix} \qquad \text{(iii)} \begin{pmatrix} 0 & -1 \\ 1 & 0 \end{pmatrix}$$

For (i) only, give the actual matrix multiplication. For all three give a sketch of the image, correctly lettered.

10 The eight operations given by the matrices in question 7 can be called the 'symmetry operations of a square', since they map the object square on to itself. They were studied in chapter 6. Are the symmetry operations of the octagon in number 9 identical with the symmetry operations of a square?

11 The symmetry operations of a square are repeated here in systematic order and the commonly used abbreviation for each operation is given. Make a table showing the matrix equivalent to each of these abbreviations. Do this very carefully as you may need it for future reference.

a) Reflection in the x axis $\qquad M_x$

b) Reflection in the y axis $\qquad M_y$

c) Reflection in $y=x$ $\qquad M_1$

d) Reflection in $y=-x$ $\qquad M_2$

e) Quarter turn about the origin $\qquad Q_1$

f) Half turn about the origin $\qquad H$

g) Three quarter turn about the origin $\qquad Q_3$

h) The identity $\qquad I$

12 A quadrilateral has vertices $(0,0)$, $(4,0)$, $(5,1)$ and $(2,2)$. Operate on it in turn with each of the eight matrices in question 11. Plot the object and each of the eight images on a single figure. You should have a closed polygon divided into 8 sections. Label each of these with one of the following: O (or I), $M_x(O)$, $M_y(O)$, $M_1(O)$, $M_2(O)$, $Q_1(O)$, $H(O)$ and $Q_3(O)$, where $M_x(O)$ means 'the image of O under the operation M_x'.

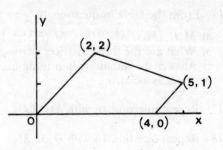

13 If A is a matrix and O is an object, $A(O)$ means 'the image of O when operated on by A'.

a) If A is $\begin{pmatrix} 3 & 2 \\ 1 & 0 \end{pmatrix}$ and O is the point $(2,3)$, what is $A(O)$?

b) If B is the matrix $\begin{pmatrix} 1 & 2 \\ 3 & 1 \end{pmatrix}$ what is $B(A(O))$? This is generally written $BA(O)$.

c) If C is $\begin{pmatrix} 2 & 1 \\ 1 & 2 \end{pmatrix}$ what is $C(BA(O))$? This is generally written $CBA(O)$.

d) If P, Q and R are matrices and W is an object, what steps would you take to find the value of $PQR(W)$?

e) If K is $\begin{pmatrix} 1 & 2 \\ 1 & 1 \end{pmatrix}$ L is $\begin{pmatrix} 2 & 1 \\ 1 & 2 \end{pmatrix}$ and M is $\begin{pmatrix} 1 & 0 \\ 2 & 1 \end{pmatrix}$

give the co-ordinates of $KLM(X)$ where X is the point $(1,2)$.

14 Here is a table for combining the symmetry operations of a square. (This table has already been studied in chapter 6.)

1st operation

		I	Q_1	H	Q_3	M_x	M_y	M_1	M_2
	I	I	Q_1	H	Q_3	M_x	M_y	M_1	M_2
	Q_1	Q_1	H	Q_3	I	M_1	M_2	M_y	M_x
	H	H	Q_3	I	Q_1	M_y	M_x	M_2	M_1
followed	Q_3	Q_3	I	Q_1	H	M_2	M_1	M_x	M_y
by	M_x	M_x	M_2	M_y	M_1	I	H	Q_3	Q_1
	M_y	M_y	M_1	M_x	M_2	H	I	Q_1	Q_3
	M_1	M_1	M_x	M_2	M_y	Q_1	Q_3	I	H
	M_2	M_2	M_y	M_1	M_x	Q_3	Q_1	H	I

a) From the table, HQ_3 (which means Q_3 first and then H) is equivalent to Q_1. What matrix represents Q_3? What matrix represents H? What is the product matrix HQ_3? Is this equivalent to Q_1?

b) From the table, what is M_1H? Write down the corresponding matrix multiplication, and check that the product is equivalent to the transformation in your first answer.

c) Repeat with Q_3M_y.

15 a) From the table, are the following operations commutative?

 i) Q_1 and H ii) Q_3 and M_x iii) M_y and M_1 iv) H and M_2

b) Write down the matrices corresponding to Q_1 and H and see if their multiplication is commutative. Repeat with each of the other pairs in *a*.

✱ **16** From the table in question 14, give the value of:

a) M_xH, $(M_xH)M_1$, HM_1, $M_x(HM_1)$. Is the product M_xHM_1 associative?

b) What are the three matrices corresponding to M_x, H and M_1?

c) Answer the questions in *a* using matrix products. Is multiplication of these matrices associative?

✱ **17** Repeat question 16 with $M_2M_yQ_3$.

✱ **18** Repeat question 16 with $Q_1M_xM_2$.

19 Every (2×2) matrix represents a transformation in the (x,y) plane. Some of these transformations can be very complicated, involving combinations of reflections, enlargements, rotations and shears. Some of these are studied in later chapters, but here are a few more examples of (2×2) matrices giving simple single transformations.

a) Operate on the arrow shown with the matrix $\begin{pmatrix} 1 & 0 \\ 0 & 0 \end{pmatrix}$

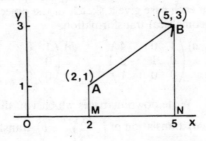

Draw a rough sketch showing the object and the image. To what geometrical transformation is this matrix equivalent? (*Hint* Draw perpendiculars AM and BN from A and B to the x axis. How is MN described in geometrical terms?)

b) Confirm your findings when the co-ordinates of A and B are:

i) (5,1) and (2,6)   ii) (3,3) and (5,5)   iii) (4,0) and (4,2)   iv) (2,0) and (4,5)

20 a) Apply the transformation $\begin{pmatrix} 0 & 0 \\ 0 & 1 \end{pmatrix}$ to the arrow in question 19.

To what geometrical transformation is the matrix equivalent? (*Hint* Draw perpendiculars to the y axis.)

b) Confirm your findings by applying the transformation to arrows given by:

i) (5,1) and (2,6)   ii) (3,3) and (5,5)   iii) (0,2) and (0,5)   iv) (0,3) and (3,1).

21 a) Apply the operation $\begin{pmatrix} 0\cdot5 & 0\cdot5 \\ 0\cdot5 & 0\cdot5 \end{pmatrix}$ to the arrow in question 19.

To what geometrical transformation is this matrix equivalent? (*Hint* Drop perpendiculars to $y=x$.)

b) Confirm your findings by applying the transformation to the arrows listed in 19b.

22 a) Operate with the matrix in question 19 on the quadrilateral $ABCD$ whose co-ordinates are (2,0), (3,3), (5,3), (4,2). Give the co-ordinates of the image. What kind of figure is the image?

b) Repeat a for the matrix in question 20.

c) Repeat a for the matrix in question 21.

d) Copy the following statement into your book and fill in the blanks: 'The images of $ABCD$ under the transformations in a, b and c are the....of $ABCD$ on ...., ...and....'

✷7D   Transformations Using (3×3) Matrices

Although every (2×2) matrix represents a transformation in the plane, not every transformation in the plane can be represented by a (2×2) matrix. Here are some examples where it is necessary to use a bigger matrix.

1 Draw the rectangle whose vertices are at (1,1), (3,1), (3,2), (1,2).
Represent the position vectors of its vertices by the matrix $\begin{pmatrix} 1 & 3 & 3 & 1 \\ 1 & 1 & 2 & 2 \\ 1 & 1 & 1 & 1 \end{pmatrix}$

(You will notice that an additional row of 1's has been written below the usual two rows.) Pre-multiply this matrix by the 'operator' matrix $\begin{pmatrix} 1 & 0 & 3 \\ 0 & 1 & 2 \\ 0 & 0 & 1 \end{pmatrix}$

and draw a rough sketch showing both object and image. Give the co-ordinates of the image, ignoring the 1's.
To what geometrical transformation is this 'operator' matrix equivalent?

2 Repeat question 1 using the operator matrices stated.
In each case give a sketch of the object and the image and state the equivalent geometrical transformation.

$$a) \begin{pmatrix} 1 & 0 & 4 \\ 0 & 1 & -2 \\ 0 & 0 & 1 \end{pmatrix} \qquad b) \begin{pmatrix} 1 & 0 & -2 \\ 0 & 1 & 1 \\ 0 & 0 & 1 \end{pmatrix} \qquad c) \begin{pmatrix} 1 & 0 & 5 \\ 0 & 1 & 0 \\ 0 & 0 & 1 \end{pmatrix}$$

3 Write down matrices which you think are equivalent to the following operations:

a) translation of $\begin{pmatrix} 2 \\ -2 \end{pmatrix}$ b) translation of $\begin{pmatrix} -1 \\ 4 \end{pmatrix}$

c) translation of 3 parallel to the x axis
d) translation of 2 parallel to the y axis
e) translation of $\begin{pmatrix} -2 \\ -4 \end{pmatrix}$

4 Using the same object as in question 1 and writing the position vectors of its vertices in the same way, find its image under the operation represented by the matrix $\begin{pmatrix} 0 & -1 & 4 \\ 1 & 0 & 2 \\ 0 & 0 & 1 \end{pmatrix}$

Draw the object and the image and try and find the geometrical transformation equivalent to the given matrix.

(*Hint* What does $\begin{pmatrix} 0 & -1 \\ 1 & 0 \end{pmatrix}$ represent?)

5 Repeat question 4 with the matrices $a) \begin{pmatrix} 0 & -1 & 6 \\ 1 & 0 & -2 \\ 0 & 0 & 1 \end{pmatrix} \qquad b) \begin{pmatrix} 0 & -1 & 7 \\ 1 & 0 & 1 \\ 0 & 0 & 1 \end{pmatrix}$

✳✳ 6 Can you find matrices equivalent to a quarter turn about:

a) (3,2)   b) (2,3)? (See hint at end of question 8.)

✳✳ 7 Can you find matrices equivalent to a half turn about:

a) (2,3)   b) (4,1)? (See hint at end of question 8.)

✳✳ 8 Can you find matrices equivalent to a three quarter turn about:

a) (2,1)   b) (2,−1)?

(*Hint for 6, 7 and 8* If the matrix you are seeking is $\begin{pmatrix} a & b & c \\ d & e & f \\ g & h & i \end{pmatrix}$

then $\begin{pmatrix} a & b \\ d & e \end{pmatrix}$ will be one of the three rotation matrices

$$\begin{pmatrix} 0 & -1 \\ 1 & 0 \end{pmatrix} \quad \begin{pmatrix} -1 & 0 \\ 0 & -1 \end{pmatrix} \quad \begin{pmatrix} 0 & 1 \\ -1 & 0 \end{pmatrix}$$

g, h and i will be 0, 0 and 1, and c and f must be found.)

✳ 9 The matrices $\begin{pmatrix} 1 & 0 & 6 \\ 0 & 1 & 4 \\ 0 & 0 & 1 \end{pmatrix}$ and $\begin{pmatrix} 1 & 0 & 2 \\ 0 & 1 & 2 \\ 0 & 0 & 1 \end{pmatrix}$ represent translations of $\begin{pmatrix} 6 \\ 4 \end{pmatrix}$ and $\begin{pmatrix} 2 \\ 2 \end{pmatrix}$ so their product should *a)* be commutative, *b)* represent a translation of $\begin{pmatrix} 8 \\ 6 \end{pmatrix}$. Check that this is so.

(See questions 1 to 3 for the representation of translations by matrices.)

✳ 10 Repeat question 9 for the matrices representing translations of

a) $\begin{pmatrix} 3 \\ 1 \end{pmatrix}$ and $\begin{pmatrix} 2 \\ 4 \end{pmatrix}$ *b)* $\begin{pmatrix} 1 \\ 0 \end{pmatrix}$ and $\begin{pmatrix} -2 \\ 3 \end{pmatrix}$.

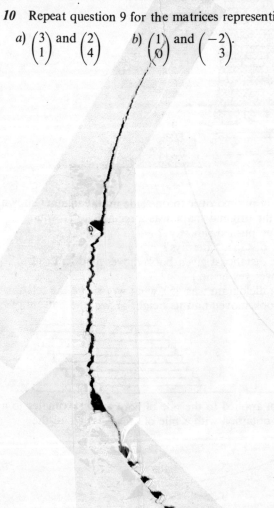

8 Shear and the Area Scale Factor of a Matrix

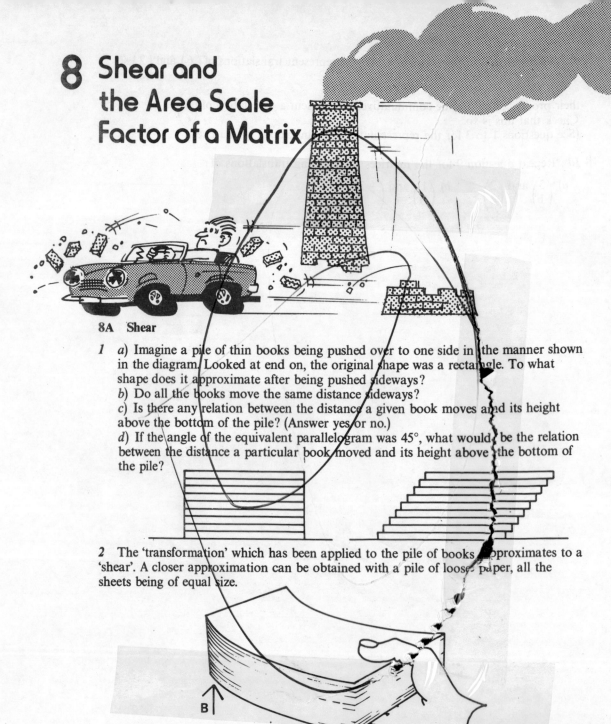

8A Shear

1 *a*) Imagine a pile of thin books being pushed over to one side in the manner shown in the diagram. Looked at end on, the original shape was a rectangle. To what shape does it approximate after being pushed sideways?

b) Do all the books move the same distance sideways?

c) Is there any relation between the distance a given book moves and its height above the bottom of the pile? (Answer yes or no.)

d) If the angle of the equivalent parallelogram was 45°, what would be the relation between the distance a particular book moved and its height above the bottom of the pile?

2 The 'transformation' which has been applied to the pile of books approximates to a 'shear'. A closer approximation can be obtained with a pile of loose paper, all the sheets being of equal size.

Carefully square the pile, grip end *A* firmly between fingers and thumb, push end *B* upwards, grip end *B* and let *A* go. The pile of paper will have been sheared. The shear can be increased by continuing to grip end *B* and pushing *A* down, then gripping *A* and letting go of *B*. A little practice will make you proficient. Apart from illustrating shearing, can you think of a practical situation in which this technique would be useful?

3 On graph paper draw a square *OABC* with vertices at (0,0), (4,0), (4,4), (0,4). Call this square the *object*. Shear it by moving *B* and *C* 8 units parallel to the *x* axis, calling them *B'* and *C'* and joining up *OAB'C'*. Call the parallelogram so formed the *image*.

a) Copy and complete the following table:

Object point	Image point	Distance moved parallel to x axis	y co-ordinate
(0,0) (0,1) (4,1) (0,2) (4,3) (0,4) (4,4) (2,4)	(2,1) (12,4)	2 units 8 units	1 4

b) Looking at the last two columns of the table, what can you say about the distance moved parallel to the *x* axis?

4 Using the same square again, move *B* and *C* through a distance of 2 units parallel to the *x* axis. Make up a similar table to that in question 3, using points of your own choice. What relation do you find this time between the distance moved parallel to the *x* axis and the *y* co-ordinate?

5 Repeat question 4, this time moving *B* and *C* through 4 units parallel to the *x* axis.

6 *a)* In question 3, every point on the *x* axis stayed unmoved or 'invariant' during the shear. The *x* axis was in fact the 'base line' or the 'invariant line' of the shear. What was the base line of the shear in question 4, and in question 5?
b) In question 3, the 'distance moved parallel to the *x* axis' was *k* times the *y* co-ordinate, where *k* had the same value for every line in the table; *k* is called the shear factor. What was the value of *k* in question 3? What was the value of *k* in each of questions 4 and 5?

7 When the base line of a shear is the *x* axis and the shear factor *k* is 2, the point (5,4) moves $2 \times 4 = 8$ units parallel to the *x* axis. In the following examples you are given the base line, the shear factor and a point. State in each case how far the given point moves, and give an arrow showing the direction of its movement.

a) *x* axis, $k=3$, (3,1)   *b)* *x* axis, $k=-3$, (2,4)
c) *y* axis, $k=1$, (3,5)   *d)* *y* axis, $k=-1$, (4,1)

Note For shears parallel to the *y* axis positive values of *k* are to be taken as giving shears as shown.

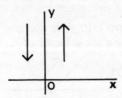

8 If the base line of a shear is $y=3$, the point (2,7) is 4 above this line, so if the shear factor is *k*, the point moves $4k$ parallel to the base line. State how far the following points move under the given shears, and give an arrow showing the direction of the movement.

a) base line $y=2$, $k=1\cdot5$, $(6,4)$ b) base line $x=3$, $k=\frac{2}{3}$, $(6,5)$

c) base line $y=-2$, $k=2$, $(1,3)$ d) base line $x=-4$, $k=\frac{1}{2}$, $(0,3)$

See note after question 7 for the sign of shears parallel to the y axis.

9 Give the co-ordinates of the vertices of the images of the following figures when they are sheared as stated. Give also a sketch of object and image and dot in the base line.

Object	Base line	k
a) (6,6), (6,2), (2,2)	$y=2$	2
b) (2,4), (4,6), (10,2)	$y=4$	$\frac{1}{2}$
c) (3,0), (6,0), (6,-4), (5,-6), (3,-2)	$y=-2$	1
d) (0,2), (2,2), (2,-2), (0,-2)	$y=0$	-2

10 Repeat question 9 using this table, in which all the shears are parallel to the y axis.

Object	Base line	k
a) (0,2), (0,6), (4,6), (4,2)	$x=0$	2
b) (6,0), (0,6), (0,0)	$x=0$	$\frac{1}{2}$
c) (0,2), (0,6), (4,6), (4,2)	$x=2$	1
d) (0,2), (0,6), (4,6), (4,2)	$x=-2$	1
e) (2,4), (4,6), (6,4), (4,2)	$x=4$	-2

See note after question 7 for the sign of shears parallel to the y axis.

✳11 Repeat question 9 using the table below. This time all the shears are parallel to the line $y=x$. Take the direction shown as positive.

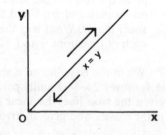

Object	Base line	k
a) (2,2), (2,4), (4,4), (4,2)	$y=x$	1
b) (2,2), (2,4), (4,4), (4,2)	$y=x$	2
c) (2,4), (4,6), (6,4), (4,2)	$y=x$	1
d) (3,6), (3,9), (0,6)	$y=x$	-1

12 Copy the figures shown into your exercise book and shear them about the base lines which are dotted. Use a shear factor of 1, and shear each figure in both directions.

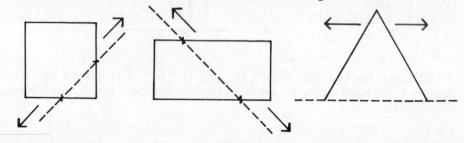

13 On squared paper draw a circle, centre the origin, radius 4 units. Using a shear factor of 1, shear it *a*) with base line $x=0$ *b*) with base line $x=-4$.

14 Whatever the base line and shear factor, *a*) a square shears into....
b) a rectangle shears into.... *c*) an equilateral triangle shears into....
d) any triangle shears into.... *e*) a circle shears into....

'3D' Shear

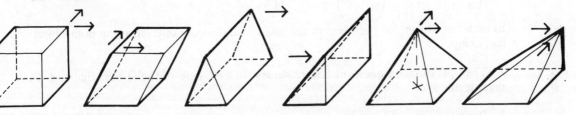

15 In the first example a cube has been sheared parallel to its base plane to give a 'parallelopiped', the 3D equivalent of a parallelogram. Its base stays square and four sides are parallelograms. It could be sheared so that all six sides were parallelograms. In the second example a triangular prism with one side an isosceles triangle has been sheared to give a prism with one side a right angled triangle. In the third diagram a pyramid with its vertex over the centre of the base has been sheared so that its vertex is over one vertex of the square base. This particular shear is used in chapter 12 to find the volume of a pyramid.

a) During a shear, the volume remains unchanged (invariant). What other properties remain invariant?
b) Name some properties that do not remain invariant.

Shear Fun

16 The figure shown in the diagram has been sheared using the dotted line as base line and a shear factor of 1.

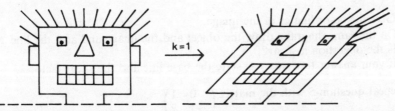

Now shear the following figures using the base line shown and the shear factor stated. In one case you are asked to use two separate shear factors. (*Hint* Work on squared paper. Select a dozen or more 'vertices' of the object figure and find their positions after shearing. Join up freehand to complete the image figure.)

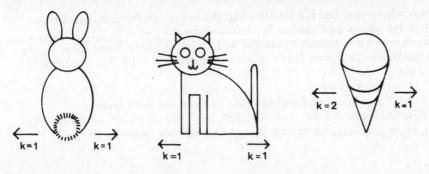

17 Shear a sheep!

8B The Shearing Matrix

1 Draw the square $OABC$ with vertices at the points (0,0), (4,0), (4,4) and (0,4). Operate on this square with the matrices

a) $\begin{pmatrix} 1 & 1 \\ 0 & 1 \end{pmatrix}$ *b)* $\begin{pmatrix} 1 & 0{\cdot}5 \\ 0 & 1 \end{pmatrix}$ *c)* $\begin{pmatrix} 1 & 2 \\ 0 & 1 \end{pmatrix}$

In each case list the co-ordinates of the image and draw a sketch showing both object and image.

2 The operations in question 1 were shears. In each case state the base line and the shear factor k.

3 Repeat question 1 using the matrices

a) $\begin{pmatrix} 1 & 0 \\ 0{\cdot}5 & 1 \end{pmatrix}$ *b)* $\begin{pmatrix} 1 & 0 \\ 2 & 1 \end{pmatrix}$ *c)* $\begin{pmatrix} 1 & 0 \\ 1 & 1 \end{pmatrix}$

All the operations are shears. In each case state the base line and the shear factor k, and give a sketch showing object and image.

(*Note* Call k positive in all three cases.)

＊ *4* Make up matrices which will perform the following shearing operations. In each case you are given the base line and the shear factor.

a) x axis, 1·5   *b)* x axis, -2 *c)* y axis, -1 *d)* y axis, 3
e) x axis, factor k *f)* y axis, factor k *g)* x axis, $k=0$

Do you recognise this last matrix?

5 Draw the square with co-ordinates (2,2), (4,2), (4,4) and (2,4). Operate on it with the matrix $\begin{pmatrix} 0{\cdot}5 & 0{\cdot}5 \\ -0{\cdot}5 & 1{\cdot}5 \end{pmatrix}$

a) Give the co-ordinates of the image.
b) On the same diagram, draw the object and the image and also the line $y=x$.
c) Is the operation a shear?
d) If your answer to c was yes, state the base line and the shear factor k.

6 Repeat question 5 with the matrix $\begin{pmatrix} 0 & 1 \\ -1 & 2 \end{pmatrix}$.

7 Draw the square with vertices at (3,3), (5,1), (7,3) and (5,5). Shear it using *a)* the matrix in question 5, *b)* the matrix in question 6. Confirm what you found about the base line and shear factors of these two matrices.

＊ *8* Any shear whose base line is a line through the origin can be represented by a (2×2) matrix, but unless the base line is $x=0$, $y=0$ or $x=y$ most of these matrices contain fractions and are difficult to operate with. However, here is one fairly simple example. Operate on the square $OABC$ whose vertices are at (0,0), (5,0), (5,5) and (0,5) with the matrix $\begin{pmatrix} 0{\cdot}6 & 0{\cdot}2 \\ -0{\cdot}8 & 1{\cdot}4 \end{pmatrix}$

Draw both object and image and find the base line and the shear factor.
(*Hint* Calling the image $OA'B'C$, join AA', BB', etc. They should all be parallel. The base line must be parallel to these and also go through one invariant point.)

* 9 Operate on the unit square $OABC$ with each of these matrices in turn and state which ones represent shears and which do not. Where they represent a shear, give the base line and shear factor.

(*Hint* Unless AA', BB', etc. come parallel, the matrix does not represent a shear.)

a) $\begin{pmatrix} 1 & 2 \\ 0 & -1 \end{pmatrix}$ b) $\begin{pmatrix} 1 & 0\cdot3 \\ 0 & 1 \end{pmatrix}$ c) $\begin{pmatrix} 1 & 0 \\ 2 & 1 \end{pmatrix}$

d) $\begin{pmatrix} 1 & 1 \\ 1 & 1 \end{pmatrix}$ e) $\begin{pmatrix} 0 & -1 \\ 1 & 0 \end{pmatrix}$ f) $\begin{pmatrix} 1\cdot5 & -0\cdot5 \\ 0\cdot5 & 0\cdot5 \end{pmatrix}$

* 10 In the (2×2) matrix $\begin{pmatrix} a & b \\ c & d \end{pmatrix}$ the number $(ad - bc)$ is called the determinant of the matrix.

Look at the six matrices in question 9 and complete this table.

	Matrix	Value of determinant $ad - bc$	Value of $a + d$	Does the matrix represent a shear?
a				
b	$\begin{pmatrix} 1 & 0\cdot3 \\ 0 & 1 \end{pmatrix}$	1	2	yes

Add a few more examples taken from this chapter or elsewhere.

* 11 Using the results of the table in question 10, answer the following questions:

a) If the determinant of the matrix is 1 and the value of $(a + d)$ is 2, does the matrix represent a shear?

b) If only one of these two conditions is satisfied, does the matrix represent a shear?

c) If neither is satisfied, does it represent a shear?

* 12 If the base line of a shear does not go through the origin, the shear cannot be represented by a (2×2) matrix. Sometimes, however, it can be represented by a (3×3) matrix. Here are a few simple examples. Write the position vectors of the vertices of the unit square as a 4×3 matrix, i.e. write them in the usual way, but with a row of 1's along the bottom:

$$\begin{pmatrix} 0 & 0 & 1 & 1 \\ 0 & 1 & 1 & 0 \\ 1 & 1 & 1 & 1 \end{pmatrix}$$

Operate on this matrix with each of the four given matrices in turn. In each case state the base line and the value of k for the resulting shear. (Ignore the row of 1's along the bottom of the image matrix.)

a) $\begin{pmatrix} 1 & 1 & -3 \\ 0 & 1 & 0 \\ 0 & 0 & 1 \end{pmatrix}$ b) $\begin{pmatrix} 1 & 0\cdot5 & -1 \\ 0 & 1 & 0 \\ 0 & 0 & 1 \end{pmatrix}$ c) $\begin{pmatrix} 1 & 0 & 0 \\ 2 & 1 & -4 \\ 0 & 0 & 1 \end{pmatrix}$ d) $\begin{pmatrix} 1 & 0 & 0 \\ 1 & 1 & 3 \\ 0 & 0 & 1 \end{pmatrix}$

8C The Area Scale Factor of a Matrix

1 Draw the quadrilateral whose vertices are at (2,5), (4,7), 6,4), and (3,3). Draw a rectangle around it as shown. Calculate:

a) the area of the rectangle,
b) the areas of the four triangles,
c) by subtraction, the area of the polygon.

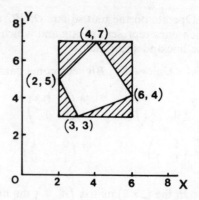

2 Repeat question 1 using the vertices (2,3), (3,6), (6,8), (8,4). This time you will have a little rectangle as well as four triangles.

Note This method of finding an area will be referred to as 'boxing in'. It gives an exact area, whereas methods based on measurement are approximate only.

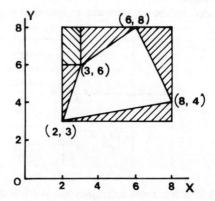

3 Draw the rectangle whose vertices are at (0,0), (2,0), (2,1) and (0,1). Operate on it with the matrix $\begin{pmatrix} 1 & 1 \\ 1 & 3 \end{pmatrix}$

Draw the image and find its area by 'boxing in'. What was the area of the object?

What is the ratio $\dfrac{\text{area of image}}{\text{area of object}}$? This ratio is the 'area scale factor of the matrix'.

4 Using the same rectangle as in question 3, operate on it in turn with each of the matrices

a) $\begin{pmatrix} 1 & 2 \\ 1 & 3 \end{pmatrix}$ b) $\begin{pmatrix} 2 & 1 \\ 1 & 2 \end{pmatrix}$ c) $\begin{pmatrix} 1 & 0 \\ 2 & 3 \end{pmatrix}$ d) $\begin{pmatrix} 1 & -1 \\ 2 & 2 \end{pmatrix}$

Enter your results in the table below, putting the results of question 3 in the first line of the table:

	Matrix	Area of object	Area of image	Area scale factor	$ad - bc$ (Δ)
3	$\begin{pmatrix} 1 & 1 \\ 1 & 3 \end{pmatrix}$	2			2
4a) etc.					

72

The last column is obtained by calling the matrix $\begin{pmatrix} a & b \\ c & d \end{pmatrix}$ and finding the value of $(ad - bc)$. Thus for the first line $(ad - bc)$ is $(1 \times 3) - (1 \times 1)$, i.e. 2.

The number $(ad - bc)$ is called the determinant of the matrix and is often denoted by the Greek letter *delta*, Δ (the Greek capital D).

5 In each of the following cases, you are given the co-ordinates of the vertices of a polygon and also a matrix. Operate on the polygon with the matrix, find the areas of object and image (by boxing in, or by any other exact method), and complete a table similar to the one in question 4.

a) (3,1), (1,3), (4,4) $\begin{pmatrix} 2 & 1 \\ 0 & 1 \end{pmatrix}$ b) (1,1), (1,4), (5,5), (7,2) $\begin{pmatrix} 2 & 1 \\ -1 & 1 \end{pmatrix}$

c) (1,0), (1,3), (3,5), (5,3), (5,0) $\begin{pmatrix} 2 & 0 \\ 0 & 2 \end{pmatrix}$ d) (0,4), (3,2), (4,4), (3,7) $\begin{pmatrix} 1 & 1 \\ -0.5 & 0.5 \end{pmatrix}$

e) (4,2), (8,2), (10,4), (7,7), (5,7), (2,4) $\begin{pmatrix} 2 & -2 \\ 1 & 1 \end{pmatrix}$

f) (4,1), (7,5), (3,8) $\begin{pmatrix} 1 & 2 \\ -1 & 1 \end{pmatrix}$ g) (1,2), (5,1), (3,4), (1,5) $\begin{pmatrix} 3 & 1 \\ 1 & 2 \end{pmatrix}$

h) (1,5), (4,2), (7,3), (7,5), (4,8) $\begin{pmatrix} \frac{1}{2} & \frac{1}{2} \\ -2 & 1 \end{pmatrix}$

6 Here are the 8 (2×2) matrices representing the symmetry operations of a square:

$M_x \begin{pmatrix} 1 & 0 \\ 0 & -1 \end{pmatrix}$ $M_y \begin{pmatrix} -1 & 0 \\ 0 & 1 \end{pmatrix}$ $M_1 \begin{pmatrix} 0 & 1 \\ 1 & 0 \end{pmatrix}$ $M_2 \begin{pmatrix} 0 & -1 \\ -1 & 0 \end{pmatrix}$

$Q_1 \begin{pmatrix} 0 & -1 \\ 1 & 0 \end{pmatrix}$ $H \begin{pmatrix} -1 & 0 \\ 0 & -1 \end{pmatrix}$ $Q_3 \begin{pmatrix} 0 & 1 \\ -1 & 0 \end{pmatrix}$ $I \begin{pmatrix} 1 & 0 \\ 0 & 1 \end{pmatrix}$

a) What is the value of the area scale factor for each one of these? Why?

✱ b) Why is its sign sometimes + and sometimes − ?

✱ **7** The rectangle $OABC$ is sheared into the parallelogram $OAB'C'$. The area is obviously unchanged since the area of a parallelogram is 'base × height' and both base and height are unchanged.

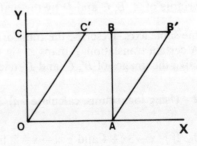

a) Is the area unchanged in every shear?

b) Is the area scale factor of a shear always 1?

c) If the determinant of a (2×2) matrix is 1, does that mean that the matrix represents a shear?

d) If the determinant of the matrix is not 1, does that mean that the matrix does not represent a shear?

Miscellaneous Examples A

A1

1 If $a=2.4\times10^6$ and $b=6\times10^3$, find the value of the following, giving your answers in standard form:

$$ab, \quad \frac{a}{b}, \quad a+10b$$

2 Given that $y=3(x^2+4z)$, write z in terms of x and y. Find the value of x when $y=6.75$ and $z=-1$.

3 Solve the simultaneous equations:

   a) $3x+y=7$ b) $3x-y=7$
       $x-y=5$ $2x+3y=12$

4 If $P=\begin{pmatrix}2 & 1\\ 0 & 3\end{pmatrix}$ $Q=\begin{pmatrix}1 & -1\\ 6 & 2\end{pmatrix}$ write down the values of:

   a) $P+Q$ b) $P-Q$ c) PQ d) QP e) P^2

5 In a bag of 24 balls, all of which are either red or green, the ratio of red and green is $2:1$. What is the probability that one ball selected at random will be red? Which part of the given information is superfluous?

6 If $f:x\to 3x+5$ and $g:x\to\frac{1}{2}(x+3)$ find:

   $f(4)$ $f(-4)$ $g(2)$ $fg(1)$ $f^2(-1)$.

Show that $f\,g(3)\neq g\,f(3)$.
Find, by drawing flowcharts if necessary, the inverse functions f^{-1} and g^{-1}. Write down the values of $f^{-1}(2)$ $g^{-1}(4)$ $f^{-1}g^{-1}(5)$.

A2

1 On graph paper draw the x and y axes for values -6 to $+6$. Plot the points $A(2,1)$ $B(4,1)$ $C(6,4)$ $D(2,6)$ and join up to form a quadrilateral. Pre-multiply the position vectors of A, B, C and D by the matrix $\begin{pmatrix}-1 & 0\\ 0 & -1\end{pmatrix}$ and draw the image $A'B'C'D'$ on the same axes. Describe the transformation brought about by this matrix.
A certain translation T maps A' on to A. What is the vector represented by T?
Give the images of B', C' and D' under the translation T.

2 Using logarithms, calculate   a) $28.6\sqrt{\dfrac{91.6}{14.2}}$ b) $\sqrt[3]{0.465}$

3 If $f:x\to 3x+4$ and $g:x\to x-5$, find $f(9)$ and $g(-1)$. Find also the composite functions fg and gf.
   Where possible write down values of x such that

   a) $gf(x)=5$ b) $fg(x)=0$ c) $fg(x)=gf(x)$

4 Rewrite each of the following to give x in terms of y:

   a) $y=3x-5$ b) $y=4(x+3)$ c) $y=\frac{1}{2}x^2$ d) $ya=bx+c$ e) $y=a(x+b)$

5 In a box of 24 chocolates, there are 9 with hard centres. What is the probability that the first one taken out is hard? If the first is hard, what is the probability that the second is also hard? In a larger box of the same kind of chocolates, how many hard ones would you expect there to be if there are 32 altogether? (Assume the same proportion of hard and soft centres.)

6 If $A=\begin{pmatrix} 2 & 1 \\ 3 & 0 \end{pmatrix}$ $B=\begin{pmatrix} 2 \\ -1 \end{pmatrix}$ $C=\begin{pmatrix} 0 & 2 & 1 \\ 3 & 0 & 1 \end{pmatrix}$, find where possible:

   a) AB b) AC c) BC d) A^2 e) B^2

A3

1 P is the point $(3,1)$ and Q is the point $(1,3)$. A transformation T maps P on to Q.

   If T is a reflection, state the mirror line.
   If T is a translation, state the vector which defines it.
   If T is a 90° rotation, find the centre of rotation.

2 A bag contains 12 red balls and 8 white ones. If I take one at random, what is the probability that it is a red one? If it is white, what is the probability that the next one taken is also white, a) if the first is replaced, b) if the first is not replaced?

3 Use logarithms to find the values of:

   a) $2 \cdot 65 \times 0 \cdot 847$ b) $\dfrac{46 \cdot 8}{0 \cdot 051}$ c) $\sqrt[3]{0 \cdot 00258}$

4 On graph paper, draw the x axis for values from -2 to 8 and the y axis for values from -4 to 4. Draw the lines $2y+x=8$ and $y-2x=-4$. From your graph find the co-ordinates of the point of intersection of the two lines. Check the accuracy of your answer by solving the equations algebraically.

5 On graph paper draw the x axis for values from 0 to 10 and the y axis for values from -6 to 4. Plot the points $O(0,0)$, $A(2,0)$, $B(2,2)$ and $C(0,2)$. Transform the square $OABC$ by pre-multiplying the position vectors of $OABC$ by the matrix S where S is $\begin{pmatrix} 1 & 3 \\ 0 & 1 \end{pmatrix}$.

Let $O'A'B'C'$ be the image and show this on your graph. Describe the transformation geometrically. What is the area of $O'A'B'C'$? Using the matrix $T=\begin{pmatrix} 1 & 1 \\ -1 & 1 \end{pmatrix}$ transform $O'A'B'C'$ into $O''A''B''C''$. What is the area of this second image? Find a single matrix which will transform $OABC$ directly into $O''A''B''C''$.

6 If $f:x \to 6-x$ and $g:x \to \frac{1}{2}(x+3)$, find the values of $f(2)$, $g(2)$ and $gf(-3)$. Write down f^{-1} and g^{-1}. Hence find the values of $gf^{-1}(0)$, $g^{-1}f(0)$ and $g^{-1}f^{-1}(7)$.

A4

1 Solve these pairs of equations:

   a) $2x-y=6$ b) $3x+y=6$
      $x+y=0$ $x-3y=7$

2 If $f:x \to 2x-3$ and $g:x \to x+4$, find the values of:

a) $f(3)$ b) $g(-5)$ c) $fg(-3)$ d) $f^2(1)$.

Show that $fg(4) \neq gf(4)$. Write down expressions for $f(a)$ and $g(a)$ and find the value of a for which $f(a)=g(a)$.

3 If $A = \begin{pmatrix} 1 & -1 \\ 0 & 3 \end{pmatrix}$ and $B = \begin{pmatrix} 4 & 6 \\ 0 & 2 \end{pmatrix}$ find:

a) $A+B$ b) $A-B$ c) AB d) BA e) A^2 f) $BA+B$
g) a matrix C such that $2A+C=B$.

4 a) If $p=q^3$ find p when $q=4$ and find q when $p=-8$.

b) Calculate the value of p when $q=0.0869$. If also $r=\dfrac{5}{q}$ find the value of r.

c) Given that $r=10$ find the value of p.

5 Let T denote the translation vector $\begin{pmatrix} -2 \\ 3 \end{pmatrix}$, M denote reflection in $y=1$ and R denote a rotation through $180°$ about $(0,0)$. P is the point $(3,0)$. Find the position of each of these images:

a) $T(P)$ b) $M(P)$ c) $R(P)$ d) $TM(P)$ e) $MT(P)$ f) $TMR(P)$ g) $MRM(P)$.

6 On graph paper draw the x axis for values from 0 to 7 and the y axis for values from 0 to 4. Draw the triangle OAB where O is $(0,0)$, A is $(3,0)$ and B is $(1,2)$.
Transform OAB by pre-multiplying the position vectors of its vertices by the matrix $\begin{pmatrix} 2 & 0 \\ 1 & 1 \end{pmatrix}$.

Draw the image of the triangle on the same axes and find the area of the two triangles. Could you have found the area of the image without drawing it first? If so, check your answer.

A5

1 Find the exact value of $370.43 \div 17$.
Use your answer to write down the values of:

a) $37.043 \div 17$ b) $370.43 \div 170$ c) $37.043 \div 0.17$

2 Use logarithms to evaluate the following:

a) $\dfrac{2.46 \times 14.8}{0.476}$ b) $\dfrac{0.00291}{0.876}$ c) $\sqrt{\dfrac{46.8}{359}}$

3 If $f:x \to 4-\dfrac{6}{x}$ find the range for the domain $\{-3, -2, -1, 1, 2, 3\}$. Draw a flowchart to help you find f^{-1}. Hence find the values of $f^{-1}(0)$ and $f^{-1}(12)$.

4 Given that $p=q+4rs$,

a) find p when $q=-2$, $r=1$ and $s=3$
b) write r in terms of p, q and s
c) find r when $p=5$, $q=3$ and $s=\frac{1}{2}$
d) if $q=s$ and $p=2s$, find r.

5 If T is the translation defined by the vector $\begin{pmatrix} 2 \\ 1 \end{pmatrix}$, R denotes a rotation through $90°$ clockwise about (0,0) and M denotes reflection in the mirror line $x=1$, describe the transformations T^{-1}, R^{-1}, M^{-1}. Find the position of each of these images if P is the point (0,2):

a) $T(P)$ b) $R(P)$ c) $M(P)$ d) $T^{-1}(P)$ e) $R^{-1}(P)$ f) $M^{-1}(P)$
g) $MT(P)$ h) $MTR(P)$.

6 On graph paper, draw the x axis for values from 0 to 8 and the y axis for values from -5 to 4. Plot the points $A(1,1)$, $B(2,1)$, $C(2,3)$ and $D(1,3)$ and join up.
Transform $ABCD$ into $A'B'C'D'$ by pre-multiplying the position vectors of $ABCD$ by the matrix $\begin{pmatrix} 1 & 2 \\ 0 & 1 \end{pmatrix}$. Show $A'B'C'D'$ on the same axes.

Transform $A'B'C'D'$ into $A''B''C''D''$ by the matrix $\begin{pmatrix} 1 & 0 \\ -1 & 1 \end{pmatrix}$ and show this image on the same axes.
Describe the two transformations in words. Find the areas of the two images. Find the matrix which would transform $ABCD$ directly into $A''B''C''D''$.

9 Circles

9A   The Circumference of a Circle
Finding the Value of π for Yourself

1   Working together in small groups collect
an assortment of tins and bottles of circular
section with diameters ranging from 4 cm to
30 cm. Measure the circumference of each
with a tape measure or by winding a piece
of thin thread several times round the
circumference and dividing the length of the
thread by the number of turns. If there is no
lip, measure the diameter by placing a ruler
across one end of the tin. If there is a lip,
use two rectangular blocks as shown, and
measure the distance between these blocks,
first making sure they are truly parallel.

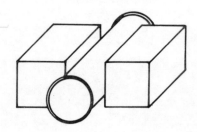

Make each measurement of circumference
and diameter at least three times and
average your results. Then plot them on a
sheet of graph paper, diameter along the
x axis and circumference along the y axis.
Draw the best straight line you can evenly
between the points and measure its gradient.
In the figure opposite the gradient would be
$\frac{AB}{BO}$. This gives an estimate of the value of π.

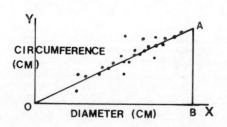

2   Try again using circular plates, discs or lids with diameters ranging from 4 cm to
30 cm. Roll each one, without slipping, along a long straight line from a marked
starting point until it has turned through one complete revolution (or three or four
revolutions for the smaller ones). Record your results on a graph as before.

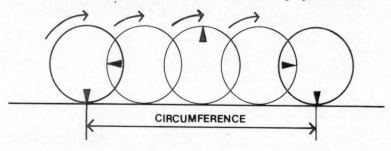

CIRCUMFERENCE

78

3 On a large sheet of paper draw a series of circles all with the same centre and with diameters 6, 8, 10...18 cm. Measure the circumference of each and record your results on a graph as before. The circumference can be measured with a piece of thin thread but this is tedious and inaccurate. If you have metal working facilities in your school workshop, you can make an opisometer and this gives an easier and more accurate measurement. The sketch shows the details. Start with the disc up tight against the handle, run it round the circumference *n* times and then run it backwards along a graduated straight line till it comes up against the handle again. The distance along the straight line is equal to *n* times the circumference (*n* should be 2, 3 or 4).

Note For questions 4 to 7 your drawing must be extremely accurate.

4 Another interesting way of estimating the value of π is to nest a circle between two regular polygons. Start with a square. On squared paper draw a square of side 12 cm and a circle of radius 6 cm touching the four sides. Join up the four points where the circle touches the square to form another square. Measure the perimeter of both squares and enter your results in the first row and the first four columns of the table in question 7 (page 80).

5 Now draw another circle of radius 6 cm and draw three diameters at angles of 60° to each other. Join up the points where these cut the circle to give a regular hexagon inside the circle. At these six points draw lines perpendicular to the radii and produce them in both directions to form a regular hexagon outside the circle. Measure the perimeters of both hexagons and record your results in the second row of the table in question 7.

6 Repeat question 5 for two regular octagons. This time start off with four diameters at angles of 45°.

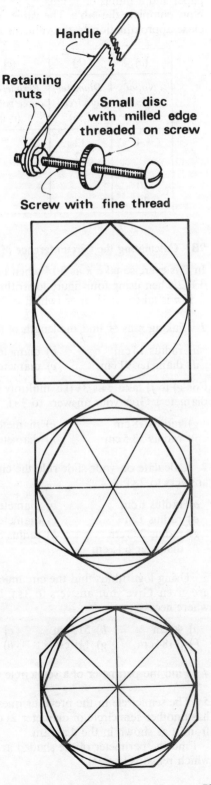

Handle

Retaining
nuts

Small disc
with milled edge
threaded on screw

Screw with fine thread

79

7 This process could be repeated again and again, but would soon become very inaccurate. You can however try it for a 10-gon and a 12-gon using a larger sheet of paper and a radius of 12 cm.

Now complete the table. The values in the last column should give an increasingly close approximation to the value of π.

(a)	(b)	(c)	(d)	(e)	(f)
No. of sides in polygon	Diameter of circle (cm)	Perimeter of outer polygon (cm)	Perimeter of inner polygon (cm)	Average perimeter $\frac{1}{2}(c+d)$	Approx value of π $(e \div b)$
4	12				
6	12				
8					
10					
12					

9B Calculating the Circumference of a Circle

In this exercise take π as 3·14 when using three-figure logarithms or slide rule and 3·142 when using four-figure logarithms.

If π is taken as $3\frac{1}{7}$ or $\frac{22}{7}$ (which is 3·1429) the answer is correct to 3 s.f.

1 Taking π as $\frac{22}{7}$ find the length of the circumference of the following circles:

a) radius 14 cm b) diameter 10·5 cm c) radius 35 cm
d) diameter 91 cm e) diameter 1 m 96 cm f) radius 63 cm

For g) to j) take π as $3\frac{1}{7}$ (i.e. multiply the diameter by 3 and add one seventh of the diameter). Give your answers to 3 s.f.

g) radius 18 cm h) diameter 12 cm i) radius 17 mm
j) radius 53·5 cm k) diameter 29·8 km

2 Calculate on your slide rule the circumference of the circles listed, giving your answers to 2 s.f.:

a) radius 6 cm b) diameter 17 cm c) diameter 27·8 cm
d) radius 1 m e) diameter $\frac{1}{4}$ km f) diameter 0·43 cm
g) radius 273 cm h) radius 81 cm i) diameter 73 mm
j) diameter 6·15 cm

3 Using logarithms find the circumference of the following circles whose diameters are given. Give your answers to 3 s.f. If using three-figure tables, round fives upwards where necessary.

a) 40 cm b) 513 cm c) 93·5 cm d) 4800 cm e) 0·763 m
f) 0·184 cm g) 119 cm h) 543 m i) 3955 km j) 217 m

4 Find the perimeter of a semicircle which has a diameter of 35 cm. Take π as $3\frac{1}{7}$.

5 The semicircle in the previous question has another semicircle of diameter 21 cm cut from it as shown in the diagram.
  Find the perimeter of the shaded area which remains $(\pi = 3\frac{1}{7})$.

6 Find the distance round a circular pond which measures 35 m across ($\pi = 3\frac{1}{7}$).

7 A donkey is tethered to a stake by a rope of length 4 m. If the donkey walks all round the stake with the rope pulled tight, how far will it have walked? ($\pi = 3.14$, answer to 3 s.f.)

8 Thread is wound on to a bobbin which has a diameter of 1·75 cm. What length of thread is there in one turn?
How many turns of the thread would be needed to wind 2·75 m on to the bobbin?
($\pi = \frac{22}{7}$)

9 Jennifer is embroidering a tray cloth to give her mother for Christmas. If she makes her piece of linen into a rectangular cloth it will measure 32 cm by 44 cm. How much edge will she have to sew?
She decides to be more ambitious and finish the edge with 'scallops'. Each one is a semicircle of radius 2 cm. Although this wastes some of her linen, Jennifer prefers it this way. How many scallops will there be on each edge? How much edge will she now have to sew?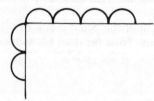

10 Using $\pi = \frac{22}{7}$, find the diameters of the following circles. You are given the length of the circumference.

a) 484 cm   b) 1331 cm   c) 77 cm
d) 242 cm   e) 3·74 cm   f) 0·66 cm
g) 11 km   h) 200 cm (answer to 3 s.f.)   i) 1 cm (answer to 3 d.p.)
j) 0·5 m (answer to 3 s.f.)

11 Using logarithms, find the radii of the following circles. You are given the length of the circumference. Give all your answers to 3 s.f.

a) 1000 cm   b) 128 cm   c) 5·4 cm   d) 8 mm   e) 940 km
f) 4·76 m   g) 2385 cm   h) 0·145 cm   i) 3720 cm   j) 45 000 km

12 A racing track is in the form of a rectangle with a semicircle on each end. If the diameter of each semicircle is 160 m, and the length of each straight side is 200 m, what is the total length of the track?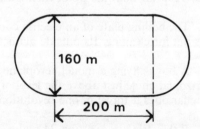

13 Repeat question 12 given that the lengths of the straight sides are 150 m and the diameters of the semicircles on the ends are 140 m. Choose the value of π which gives you an easy answer.

14 Another track of the same width as in question 13 has a total length of 1 km. What are the lengths of the straight sides?

15 Another kilometre track has a width of 175 m. What is the length of the straight sides?

16 Another track has a total length of 1200 metres and the lengths of the straight sides are 300 metres. What is the width of the track?

17 Design a racing track with a total length of 2000 metres. The width should be less than the length of the straight sides, but not less than 200 m.

18 Thin wire is being wound on to a drum. The diameter of the drum is 30 cm. How much wire is wound on in one turn? How much is wound on in 20 turns? How many turns will it take to wind on 75 metres of wire?

19 Judith has made a birthday cake for her mother's fortieth birthday. After icing, the diameter of the cake is 21 cm. She finishes it off with a wide red ribbon round the outside of the cake; this is tied with a bow. Allowing 30 cm for the bow, how much ribbon does she need altogether? (Answer to nearest 5 cm.)

20 The length of the pedal crank of a boy's bicycle is 16 cm. How far does his foot move during one turn of the pedals?

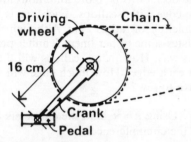

21 The roller of a 42 cm motor mower has a diameter of 18 cm.

a) How far does the mower travel forward when the roller turns through one complete revolution?
b) In a run of 24 metres (without slipping), how many times will the roller turn? Give your answer to the nearest whole number.

22 The hour hand of an alarm clock has a length of 3 cm (from centre to tip) and the minute hand has a length of 4 cm.

a) How far does the tip of each hand travel in one complete revolution?
b) How far does the tip of each hand travel in half an hour?

23 The boiling plate of an electric cooker is surrounded by a stainless steel ring which lifts out for cleaning. Its outside diameter is 20 cm. What is its circumference?

24 A boy is flying a model aeroplane on a string 20 metres long. If it is flying in a horizontal circle just above the level of his hand (which is stretched above his head) how far does it travel in one revolution?

25 If the string in question 24 makes an angle of 60° with the vertical,

a) what is the radius of the circle in which the model aeroplane is flying?
b) how far does it now travel in one revolution?

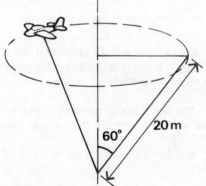

26 A 45 r.p.m. extended play record makes 340 revolutions when it is played in full. The track is of course a continuous spiral but it is convenient to think of it as 340 concentric circles. The diameter of the outer circle is then 16·8 cm and of the inner circle 11·6 cm.

    a) What is the circumference of the outer circle?
    b) What is the circumference of the inner circle?
    c) What is the ratio of these two?
    d) Could you have found this ratio without working out *a* and *b*?

27 *a*) What is the average length of the circumference of a circle from question 26?
    b) What is the approximate total length of the track (treated as 340 concentric circles)? Give your answer in km.

28 Jane's parents have bought a disused windmill and converted it into a home. The windmill is circular, tapering towards the top. Jane's bedroom is on the fifth floor and has a diameter of six metres at floor level. What is the distance round the walls at this level?

29 Jane's brother Robert has a bedroom on the fourth floor and the diameter of his room at floor level is 6·5 metres. What is the distance round the walls of his room?

30 Jane's father makes a rose bed of diameter 8 metres, with a path 60 cm wide round the outside edge. This path is to be paved with flag stones each 60 cm square. Jane works out the number needed.

    a) She finds the circumference of the inner edge of the path in cm. What should her answer be?
    b) She divides this by 60 and takes the nearest whole number. What should she get?

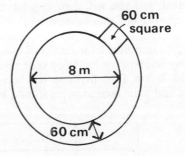

31 Robert says this will leave wedge-shaped spaces between the flag stones and it would be better to fit the flags round the outside edge of the path and trim them down to fit. So Jane does another sum.

    a) She works out in cm the diameter and circumference of the outer edge of the path. What should her answers be?
    b) She divides the circumference by 60 and takes the nearest whole number. What should her answer be?

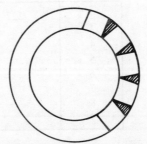

Wedge–shaped spaces

✱ **32** Their father thinks it will be too much work to trim down all these flag stones, so he decides to ask the factory to cast tapered stones. Jane works out how long the inner edge of the flag stones should be. She divides the circumference of the inner edge of the path by the number of flags needed to fit the outer edge. What should her answer be?

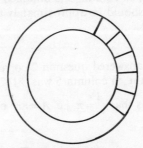

Tapered stones

83

✱ 33 Robert works out the same problem using similar triangles. He finds x in the diagram.

a) What should x be?
b) Is this the same as Jane's answer?
c) Give details of the order their father must send to the factory.

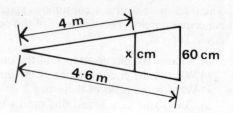

9C   The Area of a Circle
Finding Out for Yourself

1   On '2 cm' squared paper draw a circle of radius 2 cm. Do this as accurately as you can. Then find its area by 'counting squares', i.e. by counting the number of 2 mm squares inside the circle. To help you to do this, fill the circle with larger rectangles or squares. The number of 2 mm squares in each of these can be calculated, and you will then only have to count the small number of 2 mm squares left.

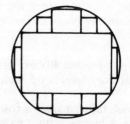

2   Repeat question 1 for circles of radii 3, 4, ... 9 cm, different members of the class working different circles. Pool your results and enter them in the first two columns of the table below. (The first row has been filled in as a guide but the values are not correct and your own answers should be used.)

1	2	3	4	5
Radius r cm	Number of 2 mm squares	Area A sq cm	r^2	$A \div r^2$
2	322	12·88	4	3·22
3				
4				
5				
6				
7				
8				
9				

Now complete the rest of the table. There are twenty-five '2 mm squares' in 1 square centimetre, so column 3 is obtained by dividing column 2 by 25. The numbers in column 5 should be approximately the same. What do you think the correct value would be?

3   If you answered question 2 correctly you will now know that the correct value of the 'constant' in column 5 was 3·142 or π.

This means that $\dfrac{A}{r^2} = \pi$, i.e. $A = \pi r^2$ or 'the area of a circle is π times the square of the radius'.

We could have deduced this result as follows:
Imagine a circle filled with a very large
number of small triangles, each with its
vertex at the centre of the circle and all
having equal bases.

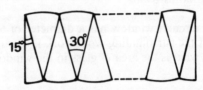

The area of one triangle is '$\frac{1}{2}$ height × base'
so, as all the triangles have the same height,
the sum of the areas of all the triangles is
'half the height × sum of the bases'.

a) What is the approximate height?
b) What is the 'sum of the bases'? Express this in terms of r. Combining these
two results shows that the sum of the areas of all the triangles is approximately
$\frac{1}{2} \times 2\pi r \times r$, i.e. the area of the circle is πr^2.

4 Now draw a circle of radius 6 cm on thin
card or stiff paper. Draw six diameters at
angles of 30°, and cut the circle into twelve
equal sectors. Divide one sector into two
along the median line, and stick the sectors
on a sheet of paper to form a rough
rectangle as shown in the diagram.

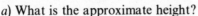

Estimate the length and breadth of this rectangle and calculate its area. See how
closely this agrees with $\pi (6^2)$.

You might get a better result by using 20 or 24 sectors, but drawing errors soon
become serious.

✱ *5* You can also investigate the area of a circle by nesting it between two regular
polygons as in **9A** 5, 6 and 7. Find the area of each polygon by measuring the height
and base of small triangles, and construct a table similar to the one used in the
questions quoted. This time divide the mean area by r^2 and the final column should
contain numbers which approximate more and more closely to π as the number or
sides in the polygon increases. (Each member of the class could do one polygon and
the results could be pooled.)

9D The Area of a Circle

1 We have seen that the area of a circle is πr^2 where r is the radius. Now calculate
the areas of the following circles whose radii are given. Take π as $\frac{22}{7}$.

a) 7 cm   *b)* 2·8 cm   *c)* 10·5 cm   *d)* 2·1 m
e) 1·4 cm   *f)* 21 cm   *g)* 140 cm   *h)* 0·7 cm

For *i* to *m* take π as $3\frac{1}{7}$, i.e. multiply the (radius)^2 by 3 and add one seventh of the
(radius)^2. Give your answers to 3 s.f.

i) 11 cm   *j)* 20 cm   *k)* 30 cm   *l)* 8 cm   *m)* 100 cm

2 Find the area of the following circles whose radii are given. Take π as 3·142, use
logarithms and give your answers to 3 s.f.

(*Note* If you use 3-figure tables, use $\pi = 3·14$ and remember the 3rd s.f. is doubtful.)

a) 6·2 cm   *b)* 13 cm   *c)* 19 cm   *d)* 220 cm   *e)* 3·5 m
f) 0·5 km   *g)* 8 mm   *h)* 137 cm   *i)* 89 cm   *j)* 15 cm

3 Check your answers to question 2 using your slide rule. You can only be certain of 2 s.f. For which questions would this have been sufficiently accurate?

4 Work rough answers to question 2, taking π as 3·0, approximating the radius to 1 s.f., and rounding fives upward where necessary.

5 Find the area of a place mat which is in the form of a circle of diameter 22 cm.

6 A circular rug with a diameter of 3 m is laid on a rectangular floor which measures 4 m by 5 m. What area of floor is uncovered?

7 *a)* Which has the larger area, a semicircle of radius 8 cm or a circle of radius 6 cm?
b) Would a semicircle of radius 8·5 cm have a larger or smaller area than a circle of radius 6 cm?

8 A circular window has a diameter of 60 cm. If the surrounding woodwork is 5 cm wide, what is the area of glass in the window?

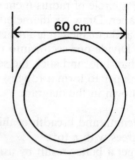

9 On the approach road to a motorway a roundabout has a diameter of 100 metres. What is its area in hectares? (A hectare is 10 000 square metres. The answer can be worked in your head.)

10 A lady buys a basketwork dog-bed for her Great Dane. The bottom of the bed is circular and has a diameter of 0·9 metres. What is its area?

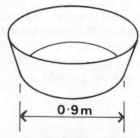

11 The circular base of a plastic flowerpot has a diameter of 8 cm. There are ten drainage holes in the base, each of approximate diameter 0·5 cm.

a) What is the area of the base of the pot?
b) What is the combined area of the ten drainage holes?
c) What fraction of the area of the base is taken up with drainage holes?

Give your answers to *a* and *b* in terms of π. Do not work them out.

12 You are given the areas of ten circles. Use your slide rule to find the radius of each, taking π as 3·14 and giving your answer to 2 s.f.

    a) 90 cm²   *b)* 16 cm²   *c)* 1 square metre   *d)* 3·5 cm²   *e)* 220 cm²
    f) 1000 cm²   *g)* 10 000 cm³   *h)* 71 cm²   *i)* 710 cm²   *j)* 0·5 cm²

Do a rough check on your first five answers. Take π as 3 and use the radius you found (abbreviated to 1 s.f.) to calculate the area. This should agree roughly with the given area.

86

13 You are given the areas of 10 circles, all areas being in cm² unless otherwise stated. Taking π as 3·142 and using logarithms, find the radius to 3 s.f. (If using three-figure tables, take π as 3·14 and remember that the third significant figure will be unreliable.) Do a rough check for the first five, using the method of question 12.

a) 84   b) 103   c) 67   d) 670   e) 22·5
f) 225   g) 0·85 m²   h) 1 m²   i) 56·2   j) 3·88

14 If A is the area of a circle and r the radius, we know that $A = \pi r^2$.

a) Rewrite this formula putting $\frac{22}{7}$ for π.
b) Make r the subject of the new formula.
c) Using the result of b find the radius of each of the following circles:

i) area 154 cm²   ii) area 38·5 cm²   iii) area 616 cm²
iv) area 1386 cm²   v) area 346·5 cm²

15 You are given the circumferences of 6 circles. Find their areas. Give your answers to 3 s.f. Take π as 3·142 (or 3·14 if you use three-figure tables).

a) 120 cm   b) 500 cm   c) 3·2 cm   d) 7·5 cm   e) 61 cm   f) 14 cm

16 Semicircles are drawn on the outside of the shorter sides of a rectangle measuring 8 cm by 6 cm. What is the total area enclosed? If the semi-circles were drawn on the inside of the shorter sides, what would be the area enclosed between the long sides and the semicircles?

17 Find in hectares the area of the race track shown. Give your answer correct to 3 s.f.

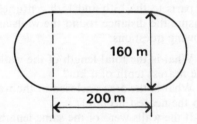

160 m

200 m

18 Jane lives in a converted windmill. Her bedroom is on the fifth floor and has a diameter of 6 metres at floor level. Find the floor area, ignoring any loss of area due to stairs, lift shafts, etc.

19 If there is a circular carpet in the middle of the floor of Jane's bedroom, leaving a polished surround 0·75 m wide, what is the area of this carpet? (Give your answer to the nearest tenth of 1 m².)

20 In Jane's bedroom there is a nest of occasional tables with tops which are circular, the diameter of the smallest being 30 cm and of the largest 70 cm. Find the area of each of these.

21 Jane's brother Robert has a bedroom one floor lower than Jane, and its diameter at floor level is 6·5 m. Most evenings he moves the furniture against the walls and sets up his model railway in the middle of the room. If no piece of furniture protrudes more than 1 metre from the wall, what is the outer radius of the largest track he can assemble? What is the length of the outside circumference of this track? What is the area it encloses?

22 When special friends of Jane's mother come to tea, she gets out the best china. The tops of the cups have a diameter of 8 cm. What is the area? The plates have a diameter of 15 cm. What is their area? There are also some cut-glass sundae dishes with a top diameter of 9 cm. What is the area of the top?

23 Outside the front door of their home, Jane's father has made a rose bed whose diameter is 8 m.

a) What is its area?
b) There is a path surrounding the rose bed and this is 60 cm wide. What is the area of the circle formed by the outer boundary of this path?
c) By subtracting the area in *a* from the area in *b* find the area of the path.
d) If it is to be laid with flag stones which are each 60 cm square, what is the area of one flag stone and how many will be required altogether?
e) If you worked **9B** 30 and 31, should your answer agree with either of those answers? Discuss.

24 Jane and Robert have a pet nanny goat called Sue. They take it in turns to milk Sue each morning before leaving for school. The seat of their three-legged milking stool has a diameter of 25 cm. What is its area?

25 Sue is tethered by a chain 10 metres long in the field near their home. The chain is fastened to an iron stake which is moved to a fresh spot each morning. What area of grass can she reach in one day?

26 Chester is the only city in England that still possesses its walls perfect in their entire circuit. Parts of the existing walls go back to Roman days soon after AD 100 and parts to the 13th and 15th centuries. The map shows the walls as they are today. Measure the distance round the walls as accurately as you can and answer the following questions:

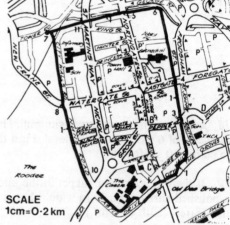

a) What is the total length of the walls to the nearest tenth of a km?
b) What is the area enclosed in the walls (to the nearest hectare)?
c) If the walls were of the same length, but circular in plan, what would be the diameter of the circle (to the nearest tenth of a km)?
d) What area would they then enclose (to the nearest hectare)?

(*The information and map are taken from the Official Guide to Chester and reproduced by permission of the Chester City Council.*)

SCALE
1cm=0·2 km

9E

Lengths of Arcs

1 Each of the two diagrams shows an arc of a circle, marked with a heavy line, and the two bounding radii.

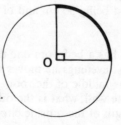
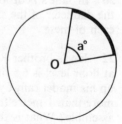

a) What fraction of the circumference is the first arc?
What fraction of the circumference is the second arc if

b) $a=60°$ *c*) $a=45°$ *d*) $a=30°$
e) $a=150°$ *f*) $a=200°$ *g*) $a=x°$

2 Find the lengths of the following arcs. In each case you are given the radius of the circle in cm and the angle between the two bounding radii. Give your answers to 3 s.f.

   a) 6, 30°   *b*) 7, 45°   *c*) 10, 90°   *d*) 4, 120°   *e*) 5, 71°   *f*) 11, 39°

3 *a*) An arc is $\frac{1}{6}$ of the circumference. What is the angle between the bounding radii? (This angle is also known as the angle *subtended* at the centre by the arc.) Repeat question *a* if the arc is *b*) $\frac{1}{4}$ *c*) $\frac{1}{8}$ *d*) $\frac{3}{8}$ *e*) $\frac{1}{5}$ *f*) 0·35   *g*) $\frac{2}{7}$
h) 0·17   *i*) $\frac{3}{11}$ of the circumference. Give your answers to *g*, *h* and *i* to the nearest degree.

4 You are given the radius of a circle and the length of an arc of the circle. Find the angle subtended at the centre by the arc. All lengths are in the same unit.

   a) 7, 5·5   *b*) 14, 11   *c*) 9, 3   *d*) 35, 22   *e*) 21, 33

Sectors

5 The area shown shaded in the figure is known as a *sector* of a circle. If the angle of the sector is 180°, what does the sector become?

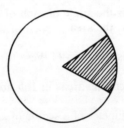

6 If the two radii bounding a sector are at an angle of 60° to each other,

   a) express the area of the sector as a fraction of the area of the circle
   b) express the length of the arc of the sector as a fraction of the circumference.

7 Repeat question 6 for angles of

   a) 90°   *b*) 20°   *c*) 30°   *d*) 120°   *e*) 48°   *f*) 103°   *g*) 29°

8 You are given the radius of a circle and the angle of a sector. In each case find the area of the sector and the length of the arc. Where your answer does not come to a whole number or simple fraction, give it to 3 s.f.

   a) 6 cm, 60°   *b*) 10 cm, 90°   *c*) 8 cm, 45°
   d) 15 cm, 36°   *e*) 11 cm, 77°   *f*) 12 cm, 86°

9 The arc of a sector of a circle of radius 7 cm is 4 cm long.

   a) What is the circumference of the circle in terms of π?
   b) What fraction of the circumference is the arc?
   c) What is the area of the circle (again in terms of π)?
   d) What is the area of the sector?

Note Leave π as a symbol throughout your calculation. It will eventually cancel out.

10 Repeat question 9 with an arc of length 8 cm and a radius of 6 cm.

11 You should now be able to see that the area of a sector is 'half the length of the arc of the sector × the radius'. Could you have deduced this in another way? If so, explain how.

12 You are given the length of the arc of a sector and the radius of the circle (both in cm). In each case find the area of the sector and the angle at the centre. Give areas to 3 s.f. where appropriate and angles to the nearest degree. Is one of the sets of information given impossible? Discuss.

a) 6, 8	*b)* 15, 10	*c)* 12, 9	*d)* 17·5, 17·5	*e)* 10, 4
f) 4, 10	*g)* 2, 18	*h)* 18, 2	*i)* 2π, 6	*j)* 31·4, 10

13 Find *a)* the perimeter, *b)* the area of a quadrant of a circle of radius 14 cm. Take π as $3\frac{1}{7}$.

14 Find the perimeter of a sector of a circle of radius 5 cm, the bounding radii making an angle of 60° with each other.

15 A box of cream cheese contains 8 pieces cut into the shape of sectors of a circle of diameter 10·5 cm. Find *a)* the area, *b)* the perimeter of the top of each section.

16 A pie chart is used to represent three quantities in the ratio of 2:3:4.

a) What is the angle of each of the three sectors?
b) What is the area of each of the three sectors, if the radius of the circle is 10 cm?
c) What is the ratio of these three areas?

✳*17* Why were the areas in 16*b* not in the ratio $2^2 : 3^3 : 4^2$?

✳*18* The perimeter of a sector of angle 60° is 6 cm.

a) Calling the radius *r*, write down an expression for the perimeter in terms of π and *r*. Get an equation for *r* by making this expression equal to 6.
b) Solve the equation for *r* taking π as 3.
c) Repeat *b* taking π as 3·14.

✳*19* The perimeter of a sector of angle 40° is 12 cm. Find the radius.

9F Miscellaneous

1 The diameter of a bicycle wheel is 50·8 cm when the tyre is fully inflated.

a) How far forward does the bicycle travel when the wheel makes one complete revolution? Give your answer in metres to 3 s.f.
b) How many revolutions does the wheel make when the bicycle travels 0·5 km? Give your answer correct to the nearest whole revolution.

2 The length of the hour hand of a clock on the steeple of a country church is 120 cm and the length of the minute hand is 180 cm.

a) How far does the tip of each hand travel in one complete revolution?
b) How far does the tip of the hour hand travel between 1000 and 1700 hours?
c) How far does the tip of the minute hand travel between 1320 and 1350 hours?

90

3   A machine operator punches six holes each of 2·1 cm diameter out of a circular disc of thin metal of diameter 14 cm. What area of metal is left? Give your answer correct to 3 s.f.

4   *a)* The angle of a sector cut out from a circle of radius 6 cm is 72°. What is its perimeter?
b) The perimeter of a sector cut out from a circle of radius 10 cm is 42 cm.
i) What is the arc length? *ii)* What is the area?

5   David wraps a tapemeasure tightly round Peter's head and finds that it reads 22 inches. If Peter's head had been truly circular, the diameter of the circle would have been $22 \div \frac{22}{7}$, i.e. 7 inches. 'And seven,' says David, 'is the size you need in hats.' Peter agrees. He wears size seven.
   Make similar measurements for yourself and your friends and see if the sizes come correct. But you must use inches. You can't use centimetres.

6   Taking the radius of the earth as 6370 km, calculate the length of the equator correct to 3 s.f.

7   A box of cream cheese has an inside diameter of 10·5 cm and contains six portions of cheese each in the shape of a sector.

   a) What is the approximate perimeter of each? (Take π as $3\frac{1}{7}$.)
   b) What is the approximate area of the top surface of each?

8   Jane and Robert are making pancakes. The bottom of Jane's pan has a diameter of 20 cm but Robert's is only 16 cm. When they have finished, their mother allows Jane to eat two of her pancakes, and Robert to eat three of his. Assuming they are of the same approximate thickness, was this fair?

9   The floor of a bell tent has a diameter of 4 metres and sleeps 8, each person having an area in the shape of a sector.

   a) What is the approximate width of the sector at its widest part? (Give the arc length as your answer.)
   b) What is the approximate area of each sector?
   c) Is this more or less than the area of a single bed which measures 2 m by 0·9 m?
   d) Express the difference as a percentage of the area of the bed.

10   A circular disc of diameter 6 cm is cut from a square of aluminium of side 8 cm. What area of aluminium remains?

11   A boy has a piece of hardboard measuring 20 cm by 60 cm.

   a) What is the largest circular disc he can cut out from this sheet?
   b) What area of hardboard remains?
   c) What 'useful area' remains?

12   His sister has a remnant of black crepe also measuring 20 cm by 60 cm. She decides to make a 'swing skirt' for one of her younger sister's dolls. The skirt is to be a complete circle, with a circular hole of diameter 5 cm in the centre for the doll's waist. She can get a circle of diameter 30 cm by cutting out two semicircles and joining them on her sewing machine.

   a) Draw a rough sketch showing how she should cut the cloth.

b) If she now sews two circles of red braid round the bottom of the skirt, 1 cm and 2 cm from the bottom, and another circle of red braid distant $1\frac{1}{2}$ cm from the top of the skirt, how much braid will she need altogether? (Ignore the loss of width due to seams and hems.)

13 Which has the larger area, a semicircular fireside mat of diameter 1·6 m or a circular mat of radius 0·6 m?

✳ 14 A widower decides to try making mince pies for his young children. He prepares the pastry and rolls it out to an even thickness in the shape of a rough circle, a little larger than the top of his preserving pan. He cuts out circles of pastry using the top of a cup, lays them in his bun tin which holds twelve pies, and then puts in the mincemeat. He gathers up the remaining pastry, folds it into a ball, rolls it out to about the same thickness as before, cuts out the tops using a slightly smaller cup and finishes off the pies. If the preserving pan has a diameter of 35 cm and the larger cup a diameter of 7 cm, how does he know that he has enough pastry, and a little over, before he starts cutting out the circles? (You can work the answer very simply using a ratio.)

✳ 15 A running track is in the form of a rectangle with a semicircle on each end. The length of the 'straight' is 600 metres and the radius of the outside semicircle on each end is 132·1 metres. Inside the outer periphery are 4 lanes, each 1·2 metres wide.

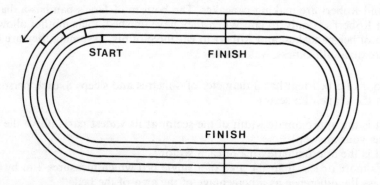

a) What is the length of the 'straight' on each of these lanes?
b) What is the radius of the inside edge of the semicircular end of each lane?
c) Give the length of each of the four lanes, measured right round the track on the inside edge. (In actual practice the length of each track is measured 30 cm from its inside edge.)

✳ 16 In a 400 metre race the starting point for the inner lane on the track described in question 15 is at the end of the straight, and the competitors start running in the direction shown. The finishing point is the same for all four tracks.

a) Where will the finishing point be?
b) Where will the starting points for the other lanes be?

✳ 17 For the track described in question 15 where would the staggered starting points be for a) an 800 metre race, b) a 1500 metre race?

* 18 A windlass is used to wind a bucket of water up a well.

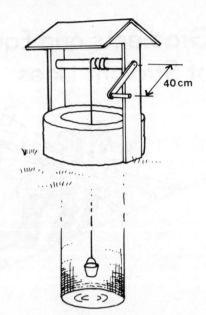

a) When the level of water is 7 metres below the top of the well, how many turns of the windlass are required to bring the bucket up to the top? The diameter of the drum of the windlass is 16 cm and the thickness of the rope is to be neglected. Answer to the nearest complete turn.

b) If the handle on the windlass is fixed to the centre of the drum by an arm which is 40 cm long, how far, to the nearest metre, does the winder's hand travel *i)* in a single turn, *ii)* in winding up the bucket?

c) You could have answered *b ii* very simply without first working out *a* or *b i*. How?

** *d) For good physicists only*
If the bucket of water weighs 12 kilograms, what steady force (in kg weight) would have to be applied to the handle to wind up the bucket? Ignore friction and the weight of the rope.

* 19 To make a circular carpet of diameter 4·5 m for Jane's bedroom her mother buys a roll of self-coloured carpet 1·5 metres wide. From this she cuts a circle of 1·5 m diameter for the centre of the carpet, and then 16 pieces in the shape of trapeziums to make the outer ring. The pieces of carpet are stitched together and firmly backed by a professional carpet fitter. The carpet is not truly circular. It is a 16-gon. But the pile spreads and it soon appears to be truly circular. Now answer the following questions.

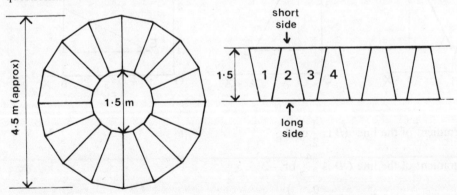

a) What is the circumference of the inner edge of the outer ring?
b) What is the circumference of the outer edge of the outer ring?
c) Remembering that there are 16 trapeziums, what is the length of the short side and of the long side of each trapezium?
d) How long a strip of carpet should Jane's mother buy? (Answer to nearest ½ metre.)

* 20 Which is travelling the faster, a bullet fired from a rifle and travelling at 50 km a minute, or a bolt 30 cm from the centre of a flywheel which is rotating at the rate of 3000 revolutions per minute? (Take π as 3.)

10 Gradients and Equations of Straight Lines

10A

Note on 'gradients'

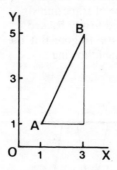 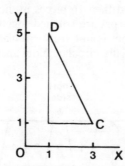 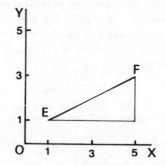

The gradient of the line AB is $\dfrac{4}{2}$ or 2

The gradient of the line CD is $\dfrac{4}{-2}$ or -2

The gradient of the line EF is $\dfrac{2}{4}$ or $\dfrac{1}{2}$

1 On the same axes draw the lines $y=x$, $y=x+1$, $y=x+3$, $y=x-1$.

   a) What do you notice about these four lines? Write down for each line *i)* its gradient, *ii)* the point where it cuts the y axis.

   b) Still using the same axes, draw the line $y=5-x$. Write down the gradient of this line and the point where it cuts the y axis.

2 The points (0,2), (1,3), (2,4), (3,5) all lie on the same straight line. You will see by looking at these co-ordinates that the x and y values increase at the same rate as one another, i.e. between two consecutive points the increase in y is the same as the increase in x. Write down a) the equation of the line, b) the gradient of the line, c) the point where the line cuts the y axis.

3 Draw the lines $y=2x$, $y=2x-1$, $y=2x+2$ all on the same axes. What do you notice about these three lines?
The points $(0,-3)$, $(1,-1)$, $(2,1)$, $(3,3)$ all lie on a straight line. Between consecutive points x increases 1 unit. What is the increase in y between consecutive points? What is the gradient of the line? Draw the line on the same axes as the other three lines and write down its equation.

4 The points (0,9), (1,6), (2,3), (3,0) all lie on a straight line. Between consecutive points the x co-ordinate increases by 1 unit. Find the corresponding change in the y co-ordinates. What is the gradient of the line? Draw a sketch of the line.

5 The points (0,0), (3,1), (6,2), (9,3) all lie on a straight line. By drawing the line or otherwise, write down the increase in the y co-ordinates between points whose x co-ordinates increase by 1 unit. What is the gradient of the line on which these points lie? What is the equation of the line?
The points (0,1), (3,2), (6,3), (9,4) all lie on a line parallel to the first line. Write down its equation in such a way that there are no fractions.

6 Find the co-ordinates of three points on the line $2y+x=0$. Plot the points accurately and draw the line.
Find two points on your line whose x co-ordinates differ by one unit. What is the corresponding change in the y co-ordinates of these two points? As the x co-ordinates increase, do the y co-ordinates increase or decrease? What is the gradient of the line? Write its equation in the form $y=$.
Write the equations of the lines parallel to this line which cut the y axis at a) (0,1), b) $(0,-2)$. If your answers to a and b have fractions, rewrite the equations without fractions.

7 The points $(-2,-3)$, (0,0), (2,3), (4,6) all lie on a straight line. Plot the points and draw the line. For an increase of 1 unit in the x co-ordinates of two points, find the corresponding change in their y co-ordinates.
What is the gradient of the line? What is its equation?
Write down the equations of lines parallel to this line which cut the y axis at a) $(0, -1)$, b) (0,1), c) (0,3).

8 A straight line passes through the two points $(2,-3)$ and $(4,-6)$. Write down the co-ordinates of the point mid-way between them. The x co-ordinates of these three points taken consecutively increase by 1 unit. What is the corresponding change in their y co-ordinates? What is the gradient of the line joining them? Where does the line cut the y axis? Write down the equation of the line.

9 For each of these lines write down the gradient and the co-ordinates of the point where the line cuts the y axis:

a) $y=3x+5$ *b*) $y=4-x$ *c*) $x+y=0$ *d*) $4y=3x$
e) $y=1-2x$ *f*) $y=2x-3$ *g*) $2y=5x$ *h*) $y=3-4x$
i) $y=6+x$ *j*) $y=4x-3$

10 Rearrange each of the following equations to the form $y=mx+c$ and then write down the gradient of each line and its point of intersection with $x=0$:

a) $2y=x-4$ b) $3y=1-x$ c) $3y=2x+7$ d) $x+3y=4$
e) $3x+2y=0$ f) $4x+y=2$ g) $2x-5y=6$ h) $4-y=3x$
i) $2x+4y=9$ j) $3x-7y=6$

11 If a straight line has the equation $y=mx+c$,

a) what is its gradient?   b) at what point does it cut the y axis?

12 In each of the following examples you are given a pair of points, one of which is on the y axis. Write down the gradient of the line joining the pair of points, and then, knowing where it cuts the y axis, write down its equation.

a) (0,1) (1,2)   b) (0,1) (1,−1)   c) (0,3) (1,6)   d) (0,−1) (1,0)
e) (0,2) (1,−3)   f) (0,−3) (1,−1)   g) (0,−2) (1,1)   h) (0,−2) (1,−4)
i) (0,4) (1,2)   j) (0,4) (1,7)

13 Find the gradient of the line joining each of these pairs of points:

a) (1,1) (2,3)   b) (0,1) (2,2)   c) (1,2) (3,5)   d) (1,−1) (2,0)
e) (3,1) (5,−1)   f) (2,0) (4,−2)   g) (0,0) (−1,−2)   h) (1,0) (0,1)
i) (2,4) (0,6)   j) (3,1) (5,0)

14 Write down the gradient of each of the following lines:

a) $y=x+7$ b) $y=5-x$ c) $2y=4-x$ d) $3y=2x+1$
e) $2y=1-4x$ f) $2y=6x-5$ g) $y=3-2x$ h) $2y+x=5$
i) $3x+2y=1$ j) $4x-2y-5=0$ k) $y=px+q$

15 State the gradient of each of the following vectors:

a) $\begin{pmatrix}3\\4\end{pmatrix}$ b) $\begin{pmatrix}7\\2\end{pmatrix}$ c) $\begin{pmatrix}4\\-3\end{pmatrix}$ d) $\begin{pmatrix}-2\\-1\end{pmatrix}$ e) $\begin{pmatrix}9\\-6\end{pmatrix}$ f) $\begin{pmatrix}1\\-2\end{pmatrix}$ g) $\begin{pmatrix}-2\\5\end{pmatrix}$ h) $\begin{pmatrix}3\\3\end{pmatrix}$

10B   Straight Line Graphs and their Gradients

1 The graph shows the distance of a motorist from his starting point over a period of two hours.

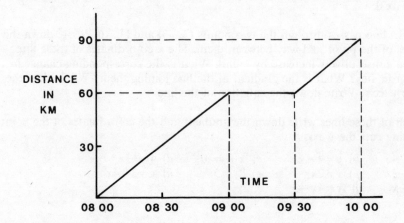

a) How far has he travelled in the first hour?

b) Since the line of the graph covering the first hour is straight, the motorist is travelling at a constant speed. What is this speed?

c) What can you say about the period between 0900 and 0930?

d) What is the speed during the last half hour?

e) If the motorist had covered the 90 km at a steady speed in the same time, what would this speed have been?

2. The graph shows the distance of a motorist from his starting point over a period of two hours.

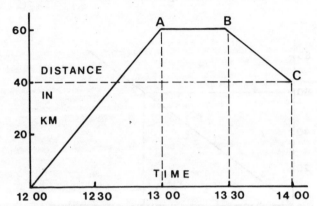

a) Explain what happened between the points marked *A* and *B*.

b) What is his speed between 1200 and 1300 hours?

c) What is happening between *B* and *C* and what is his speed during this time?

d) If the motorist wanted to arrive back at his starting point at 1445 hours, how fast would he need to travel after leaving *C*?

3 Martin and Neil were friends. Martin set out to walk to Neil's house 6 km away. He walked at a steady 6·5 km an hour. Draw a distance/time graph taking 2 cm to 10 minutes on the horizontal axis and 2 cm to 1 km on the vertical axis. From the graph find how long it took Martin to finish his walk.

At the same time that Martin set out, Neil also set out on his bicycle to visit Martin. He cycled at a speed of 18 km an hour, but after travelling for 10 minutes stopped at the library for 20 minutes. He then completed the rest of the 6 km at the same speed as before. Martin walked past the library while Neil was inside. By how many minutes did they miss one another?

If they stayed at each other's homes the same length of time before starting back again, how fast did Neil cycle if he reached home just before Martin left?

4 A slow train leaves Furzedown station travelling on an up line at 80 km per hour. Twenty minutes later a fast train leaves the same station on an up line travelling at 130 km per hour. 40 km from the station there is a stretch of single track. Find graphically how many minutes the slow train must be held back to allow the fast train through first. Use 2 cm for 10 minutes on the horizontal axis and 2 cm for 20 km on the vertical axis.

At the same time as the fast train is leaving Furzedown, another train is setting off from Oakport (a station which is 100 km from Furzedown) and travelling on the down line. If these two trains pass one another 28 minutes after starting, where are they and at what speed is the third train travelling?

How much farther will it have travelled when it meets the slow train if this train was actually held back for 10 minutes on the single track section?

5   A coach leaves a service station on a motorway with a distance of 80 km to cover. If the time is 10 o'clock and the driver estimates that he will travel at an average speed of 75 km per hour, at what time should he arrive at his destination?
Draw a graph using 2 cm to 30 minutes on the horizontal axis and 2 cm to 10 km on the vertical axis.
A car travelling in the opposite direction to the coach also sets out at 10 o'clock to cover the same 80 km. If the car has one stop of 45 minutes after passing the coach and arrives at the service station at 12 o'clock, at what speed has it travelled? At what time and where did the two vehicles pass one another?

6   The graph shows a straight line which passes through the origin. The axes are numbered.

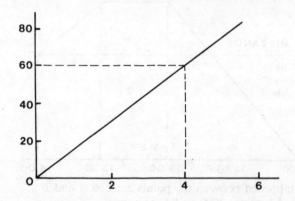

a) If the vertical axis represents cost in pence and the horizontal axis is lengths of wire measured in metres, what does the slope of the line represent?
b) If the vertical axis is distance in km and the horizontal axis is time in hours, what is the gradient of the line and what does it represent?

7   The diagram shows a straight line graph which passes through the point (0,100).

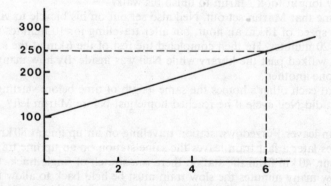

a) If the vertical axis represents the quantity of water in litres in a tank and the horizontal axis time in minutes, what is the gradient of the line and what does it measure?
b) If the cost of some photographs in pence is represented on the vertical scale and the number of photographs on the horizontal scale, what does the gradient of the line give?
c) Explain in each case why the line does not go through the origin.

8   The graph shows the velocity of a small object over a period of 6 seconds.

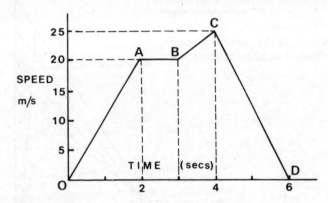

a) What is happening between *A* and *B* on the graph?
b) Between which points is the acceleration greatest?
c) What is the acceleration at the beginning?
d) What is happening between *C* and *D* and what is the rate of change of velocity over this period?

9   A small object starting from rest accelerates at a rate of 3 m per second over a period of 6 seconds. It then travels with a constant speed for 4 seconds and is brought to a halt by a constant retardation in another 2 seconds. Draw a sketch of the velocity/time graph and find the retardation in the last 2 seconds, i.e. the rate at which the velocity is reduced.

10C   Gradients of Non-linear Graphs

1   The graph shows the height of a seedling over a period of 4 weeks. Because the rate of growth is not constant, the graph is a curve.

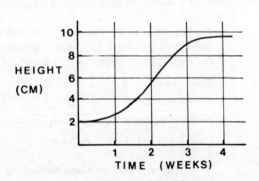

a) During which time is the rate of growth at its greatest?
b) When is it least?

2   The graphs in 10B are all straight lines which means that the gradient of each is the same over a given period. The gradient measures rate of change. The subjects shown in the graphs are listed below. In actual life, which do you think give straight lines and which would be more likely to have a varying rate of change and hence give curves?

a) Distance from starting point — time.
b) Total cost — length.
c) Quantity of water in a tank — time.
d) Total cost — number of articles.
e) Velocity (or speed) — time.

3 The accompanying diagram shows a
curve formed by 'curve stitching' with which
you are probably familiar. The more straight
lines there are, the simpler it is to fit in a
curve. The direction of the curve is changing
all the time following one line after another.

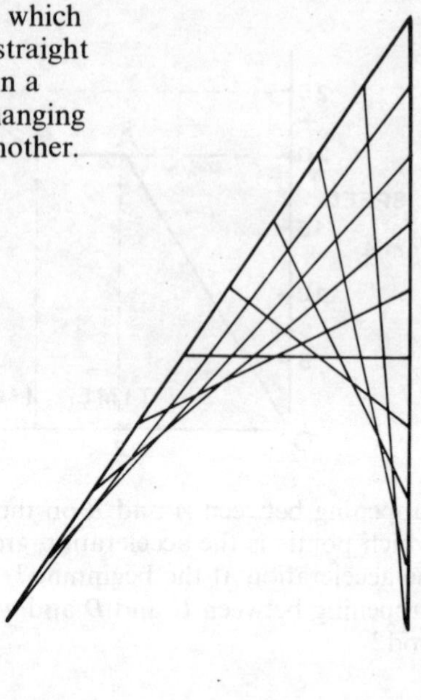

Using the scale of 2 cm to 1 unit on the x axis and 2 cm to 2 units on the y axis,
plot the following points and join up with a smooth curve.

x	-4	-3	-2	-1	0	1	2	3	4
y	0	3·5	6	7·5	8	7·5	6	3·5	0

By drawing in tangents to the curve (similar to the 'stitches') find the gradient of the
curve at the points where

a) $x=-1$ b) $x=2$ c) $x=-3$ d) $x=0$

4 Taking 2 cm to represent 1 second on the horizontal axis and 2 cm to represent
2 cm on the vertical axis, plot the following points which show the distance of a marble
from its starting point as it rolls down a sloping groove.

Time (secs):	0	1	2	3	4	5	6
Distance (cm):	0	0·25	1	2·25	4	6·25	9

Join the points with a smooth curve and find the speed of the marble after

a) 1 sec   b) 3 sec   c) 5 sec.

5 A quantity of water was heated up in a bottle closed with a rubber stopper and
then allowed to cool over a period of 18 minutes. The temperature in °C was as follows:

Time (mins):	0	1	2	3	4	5	6	8	10	12	14	16	18
Temperature:	21	24	28	32·5	34	33·8	33·6	33	32	30·8	30·1	28·8	27·2

Using 2 cm to 2 minutes on the horizontal scale and 2 cm to 2° on the vertical scale,
plot the information and draw a smooth curve. Find from your graph

a) the time interval over which the temperature is rising most rapidly,
b) the rate of increase of temperature at 3 mins,
c) the time interval over which the temperature is falling most rapidly,
d) the rate of loss of temperature at 8 mins,
e) the rate of loss at 12 mins.

6 The accompanying graph shows the distance of a train from one station over a period of two hours. Answer the following questions from the graph:

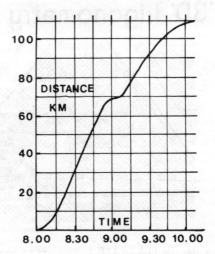

a) At what time does the train stop at the second station?

b) What was the average speed over the first half hour?

c) What was the average speed over the first 40 km?

d) What was the average speed between stations?

e) At what speed was the train travelling at 8.45?

f) What was the average speed between 9.30 and 10.00?

g) At what speed was the train travelling at 9.30?

7 A stone is thrown vertically upwards from the edge of a cliff. The table shows the distance of the stone from its starting point every half second for three seconds.

Time (secs):	0	$\frac{1}{2}$	1	$1\frac{1}{2}$	2	$2\frac{1}{2}$	3
Distance (m):	0	$3\frac{3}{4}$	5	$3\frac{3}{4}$	0	$-6\frac{1}{4}$	-15

Using 5 cm to 1 second on the horizontal axis and 2 cm to 2 m on the vertical axis, plot the points and join up with a smooth curve.

a) From your graph find the speed of the stone after $\frac{1}{2}$ sec, 1 sec, 2 sec, $2\frac{1}{2}$ sec.

b) Find also the average speeds over the first second and over the third second.

c) If after 3 seconds the stone hit the rocks at the foot of the cliff, with what velocity did it strike the rocks?

8 In a large outdoor model layout, the speed of an electric train as it moves between two stations is given by the formula $v=4t-0\cdot5t^2$ where t is the time in seconds after leaving the first station and v is the speed in metres per second.

Taking values of t from 0 to 8, work out corresponding values of v and plot them on a graph. Let 2 cm represent 1 sec on the horizontal axis and 2 cm be 1 metre per second on the vertical axis. From your graph find

a) the maximum speed of the train,

b) the acceleration of the train when $t=3$ seconds,

c) the retardation when $t=6$ seconds.

9 A car accelerates away from the kerbside. The following table gives its speed in metres per second for the first 60 seconds:

Time (secs):	0	10	20	30	40	50	60
Speed (m per sec):	0	17·1	24·2	29·7	34·3	38·3	42

Using 2 cm horizontally to represent 10 seconds and 2 cm vertically to represent 5 metres per second, draw the speed/time graph. From your graph find

a) the average acceleration over the first 10 secs

b) the acceleration 10 secs after the start ((by finding the gradient of the tangent to the curve at $t=10$)

c) the acceleration 50 secs after the start

d) the average acceleration over the last 10 seconds.

11 '3D' Trigonometry

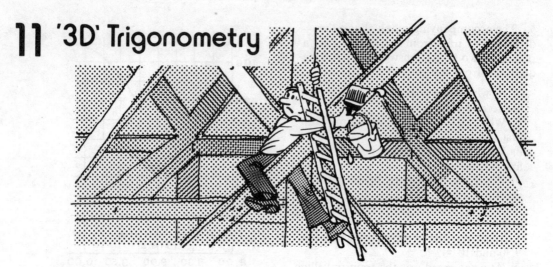

Note In 11A to 11E angles are to be calculated to the nearest degree and lengths to 3 s.f. unless otherwise stated. Where sketches of triangles are asked for, they should not be drawn to scale.

11A Pythagoras' Theorem in 3 Dimensions

1 *ABCDEFGH* is a rectangular prism of sides 3 cm, 4 cm and 12 cm.

a) Draw a sketch of triangle *ABC* (not to scale). Using Pythagoras' Theorem find the length of *AC*.

b) Draw a sketch of triangle *ACG*. Mark on the lengths of *AC* and *CG*. Using Pythagoras' Theorem, find the length of *AG*.

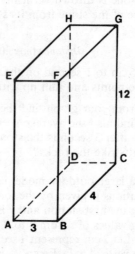

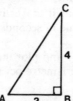

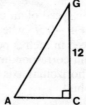

2 Using the figure from question 1, find the length of *AG*, being given the lengths of *AB*, *BC* and *CG* in that order. Give your answers to 3 s.f. All lengths are in cm. (*Hint* having found AC^2, it is not necessary to find *AC*, unless specifically asked to do so.)

   a) 4, 4, 5   *b*) 6, 8, 10   *c*) 3, 4, 7   *d*) 7, 7, 7   *e*) 7, 8, 9

3 Since $AC^2 = AB^2 + BC^2$ and $AG^2 = AC^2 + CG^2$ it follows that $AG^2 = AB^2 + BC^2 + CG^2$

This is sometimes called 'Pythagoras' theorem in 3 dimensions'. Using this method calculate the length of the long diagonal in each of the following prisms. You are given the lengths of the three sides that meet at any one vertex:

   a) 4, 4, 2   *b*) 6, 3, 2   *c*) 6, 6, 3
   d) 8, 4, 1   *e*) 9, 6, 2   *f*) 7, 6, 6
   g) 11, 10, 2 *h*) 10, 11, 12 *i*) 6, 9, 12
   j) 4, 7, 11

4 The diagram shows a square based pyramid with the vertex V directly above O, the intersection of the diagonals of the base.

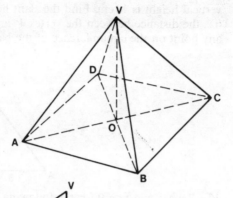

a) Draw triangle ABC. If $AB=4$, what is the length of AC? What is the length of AO?

b) Draw triangle AVO. If $OV=4$, using the length you calculated for AO, what is the length of AV?

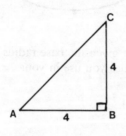

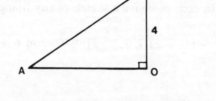

5 Calculate the lengths of the sloping edges of the following square pyramids. You are given the length of one side of the base and the vertical height, in that order. All lengths are in cm.

a) 4, 3   b) 5, 2   c) 6, 6   d) 7, 5   e) 8, 6

6 In the following pyramids, the base is a rectangle and the vertex is vertically above O, the intersection of the diagonals. You are given the lengths of the sides of the base and the vertical height of the pyramid. All measurements are in cm.
Find in each case the length of the sloping edge of the pyramid. Draw a sketch (not to scale) of every triangle you use in the calculation.

a) 3, 4, 4   b) 5, 12, 8   c) 6, 8, 7   d) 4, 6, 8   e) 5, 7, 4

7 In a square based pyramid the vertex V is directly above the centre of the base. In each case below you are given the length of one side of the base and the vertical height. Find the lengths of OE and VE where E is the mid-point of one side of the base. Draw sketches of any triangles you use.

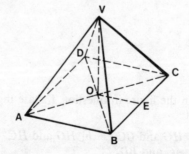

a) 4, 6   b) 6, 7   c) 2, 5   d) 8, 6

All measurements are in cm.

8 The base of a pyramid is a rectangle, and the vertex V is directly above O the centre of the base. E and F are the mid points of the shorter and longer sides of the base. You are given the two sides of the base and the vertical height, all lengths being in cm.
Find the lengths of VE and VF. Draw sketches of any triangles you use in your calculations.

a) 6, 8, 8   b) 4, 6, 8   c) 2, 4, 6   d) 6, 8, 10

9 The base radius of a cone is 8 cm and the
vertical height is 10 cm. Find the slant height
(i.e. the distance between the vertex V and
any point on the circumference of the base).

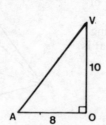
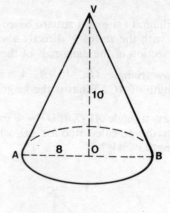

10 Repeat question 9 for the following cones. You are given the base radius and the
vertical height. In each case give a sketch of any triangles you use in your
calculations.

a) 3, 4   b) 5, 12   c) 7, 7   d) 4, 9   e) 6, 11

11B Angles between Lines

If two lines meet, the angle between them can be calculated in one of the usual ways,
e.g. by selecting a suitable triangle and using trigonometry. If two lines do not meet and
and are not parallel, they are said to be skew lines. The angle between them can then
be found by translating one line parallel to itself until they do meet.

1 Figure $ABCDEFGH$ is a cube. State
whether each of the following pairs of lines
meet, are parallel or are skew:

a) HG, HF b) HG, DE c) HG, AC
d) HF, HA e) AF, EF f) HA, GB
g) BH, HE h) GE, AB

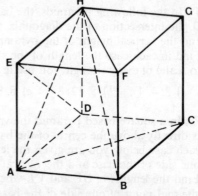

2 In the figure in question 1, state the angle between the following pairs of lines that
meet:

a) HG and GC b) HG and HC c) BA and BF d) AB and AC
e) AC and BD

3 In the figure in question 1, to find the angle between BC and AE translate BC
parallel to itself to AD. The required angle is then the angle between AD and AE, i.e.
angle EAD which is 90°.
By translating AB parallel to itself find the angle between

a) AB and HD b) AB and HE c) AB and HF

In each case state the position to which AB is translated, which angle you then
measure and its size in degrees.

104

4 Draw a sketch of triangle *AHC* in the figure in question 1. What kind of triangle is it? What is the angle between *AH* and *AC*?

5 The figure shows a rectangular prism *ABCDEFGH* with sides 3, 4, 5 cm. To find the angle between *AH* and *AD* draw the triangle *AHD*. The angle required is the angle whose tangent is $\frac{5}{3}$, i.e. 1·667. This is 59° to the nearest degree.

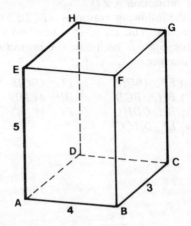

Now find the angle between the following pairs of lines. In those cases where it is necessary to consider a triangle, give a sketch of this triangle with its vertices lettered and any known lengths or angles clearly marked.

   a) AF, FB   *b) AC, AB*   *c) BC, BD*
   d) GH, GD   *e) BH, BD*

6 Using the figure in question 5, find the angle between the following pairs of skew lines. In each case state which line you are translating, and to what position it is translated. If you have to consider a triangle, show it in the manner described in question 5.

   a) AB, DG   *b) BC, AH*   *c) HG, BF*   *d) ED, GC*   *e) AC, HF*

(*Hint for e* Calculate the half angle first.)

✱ 7 *VABCD* is a square based pyramid with the vertex *V* directly above *O* the centre of the base. *AB*=4 cm, *VO*=5 cm. Find the angles between

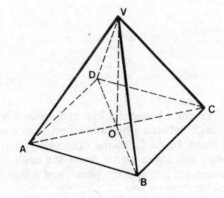

   a) AB and *VO*   *b) AB* and *BD*
   c) VA and *AB*   *d) AB* and *VD*
   e) VA and *VB*   *f) VA* and *AC*

✱✱ 8 Repeat question 7 with *AB*=4 cm, *AD*=3 cm and *VO*=5 cm.

11C The Angle between a Line and a Plane

1 The projection of a line on a plane.
To find the projection of a line *XY* on a plane, drop perpendiculars *AP* and *BQ* on to the plane from any two points *A* and *B* on the line. The line through *PQ* (extended in both directions) is the projection of *XY* on the plane.

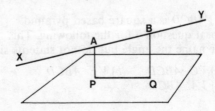

If the line XY cuts the plane at Z, then Z can be taken as one of the two points and the projection is ZQ.
In the following examples $ABCDEFGH$ is a cube. You are given a line and a plane. State in each case the projection of the line on the plane:

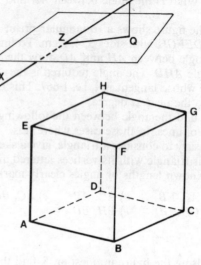

a) $EF, ABCD$ b) $EG, ABCD$
c) $BH, ABCD$ d) $BH, AEHD$
e) $BH, CDHG$ f) $FC, AEHD$
g) $EF, DHGC$

2 Repeat question 1 for the lines and planes in the square based pyramid shown:

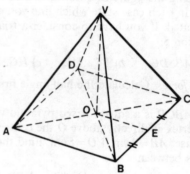

a) $VB, ABCD$ b) $VD, ABCD$
c) $VE, ABCD$

3 The angle between a line and a plane is the angle between a line and its projection on the plane. In the following examples you are given a line and a plane. Name the angle between the line and the plane, and where possible state its size.
$ABCDEFGH$ is a cube.

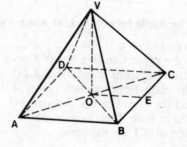

a) $ED, ABCD$ b) $HC, ABCD$
c) $AC, BCGF$ d) $AG, ABCD$
e) $CE, EFGH$ f) $FD, ABFE$

4 $VABCD$ is a square based pyramid. Repeat question 3 for the following. This time name the angle but do not state its size.

a) $VC, ABCD$ b) $VE, ABCD$
c) $VA, ABCD$

5 In this question you are given a line and a plane. Name the angle between the line and the plane. By drawing suitable triangles, calculate the size of the named angle.

a) AH, ABCD   b) AH, HGCD
c) AG, ABCD   d) AG, BFGC
e) FH, FGCB

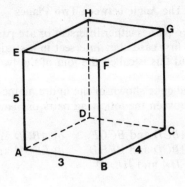

6 Repeat question 5 for the following:

a) AV, ABCD   b) EV, ABCD

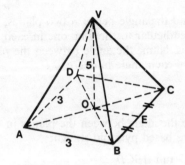

7 For the cone shown in the diagram, find the angle between the slant height AV and the base. Find also the angle between the slant height AV and the axis VO. This is called the 'semi-vertical angle' of the cone.

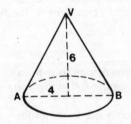

✱ 8 There is a distinction between the projection of an infinite line and the projection of a line segment. In the figure the infinite line through XY projects into the infinite line through RS, but the line segment AB projects into the line segment PQ. State the projections of the following:

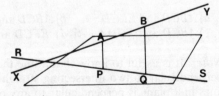

a) the line segment BC on to the plane AEHD
b) the infinite line through AB on to the infinite plane HGCD
c) the infinite line AG on to the infinite plane BFGC
d) the line segment AC on to the plane EFGH

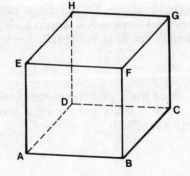

11D The Angle between Two Planes

1 Two planes either intersect or are parallel. In the first case, they intersect in a straight line and this is called the join of the two planes.

For the cube shown in the figure, name the join between the following pairs of planes:

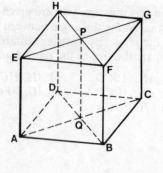

a) *ABCD* and *BCGF*   b) *ABCD* and *DCFE*
c) *ABCD* and *BFHD*   d) *EFGH* and *HFBD*
✱ e) *FHA* and *HEAD*

2 To find the angle between two planes, select a point *P* on the join. Draw *PQ* and *PR* perpendicular to the join, one in each plane. The angle *QPR* is the angle between the planes. Name the angle between the planes in question *1a* to *d* (not *e*). State also the size of each angle in degrees.

3 Name the angle between the planes in the square based pyramid shown:

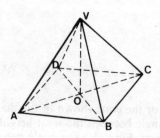

a) *VBC* and *ABCD*
b) *VBA* and *ABCD* (for this angle you will have to add extra lines to the figure)

4 Using the dimensions shown, calculate the angle between the following pairs of planes.

Where you need to use triangles for your calculation, draw the triangles, letter them and mark any known dimensions.

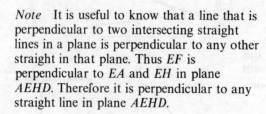

a) *ABCD* and *EFCD*   b) *ABCD* and *BFHD*
c) *ABCD* and *FGDA*   ✱ d) *EFCD* and *AEHD*

Note It is useful to know that a line that is perpendicular to two intersecting straight lines in a plane is perpendicular to any other straight in that plane. Thus *EF* is perpendicular to *EA* and *EH* in plane *AEHD*. Therefore it is perpendicular to any straight line in plane *AEHD*.

5 Find the angle between the planes *VCB* and *ABCD*. State also the angle between *VAB* and *ABCD*.

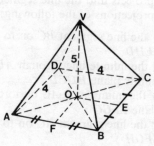

6 Repeat question 5 with $AD = BC = 3$ and the other given lengths unchanged.

11E 3D Co-ordinates

1 The diagram shows a set of co-ordinate
axes in 3D space. The point P has the
co-ordinates (x, y, z). OX, OY and OZ are the
positive directions of the axes and OX',
OY' and OZ' are the negative directions.
Draw freehand a similar set of axes and
mark on the following points:

(3,4,5) (−3,4,5) (−3,−4,5)
(3,−4,5) (3,4,−5)

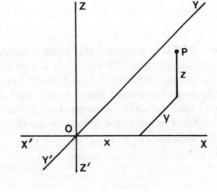

2 The diagram shows the $(+ + +)$ section
of the 3D space. Answer the following
questions:

a) What is the length of ON?
b) What is the length of OP?
c) Name the angle between OP and the
plane OXY and calculate its size.

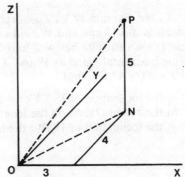

3 Repeat question 2 for the point (5,12,10).

4 Repeat question 2 for the point (4,7,4).

5 Repeat question 2 for the point (7,6,6).

6 The point (9,2,6) lies on the surface of a sphere whose centre is at O the origin.
Calculate the radius of the sphere.

7 Repeat question 6 for the points

a) (1,4,8)   *b*) (6,3,6)   *c*) (11,2,10)   *d*) (4,5,6)   *e*) (7,8,5)

11F Miscellaneous

Angles are to be calculated as accurately as your tables allow and lengths to 3 s.f.
unless otherwise stated.

1 $VABCD$ is a pyramid on a square base $ABCD$ with the vertex V vertically
above A. If $AB = 10$ cm and $VA = 12$ cm, find the lengths of the slant edges and the
angle each makes with $ABCD$.

2 $VKLMN$ is a pyramid on a rectangular base $KLMN$ in which $KL = 9$ cm,
$KN = 12$ cm. The vertex V is at a height of 15 cm vertically above K. Find the
lengths of the slant edges and the angle each makes with the base.

3 Two points X and Y are in the same horizontal plane as the foot of a tree. X is 25 m due E of the tree and Y is 30 m due S. Looking at the top of the tree from X, the angle of elevation is 45°. What is the height of the tree? What will be the angle of elevation of the top of the tree from Y?

4 $ABCD$ is a rectangular plane where $AB = 12$ cm and $AD = 8$ cm. If E and F are the points vertically below D and C and in the same horizontal plane as AB, and $DE = 4$ cm, find the angle between $ABCD$ and $AEFB$. Find also the length of the diagonal AC and the angle it makes with the horizontal.

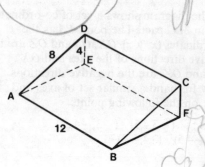

5 In the diagram $WXYZ$ is a rectangle in which $WX = 16$ cm, and $WZ = 13$ cm. If P and Q are vertically below Z and Y in the same horizontal plane as W and X such that $WP = XQ = 12$ cm, find

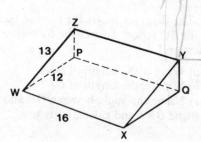

    a) the inclination of $WXYZ$ to the horizontal,
    b) the height QY,   *c)* the length WQ,
    d) the inclination of WY to the horizontal.

✱6 $PQRS$ is a square of side 16 cm. If the square is held in such a way that PQ is horizontal and $PQRS$ is inclined at 40° to the horizontal, find the angle between PR and the horizontal.

7 $VABC$ is a pyramid on a triangular base in which $AB = AC = 30$ cm, and $BC = 36$ cm. D is the mid-point of BC and O is the point in AD such that $DO = \frac{1}{3}AD$. V is vertically above O at a height of 16 cm. Find *a)* AD, *b)* the angle between VA and the base, *c)* OB, *d)* VB, *e)* the angle between VB and the base.

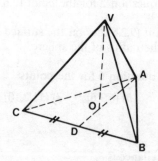

8 $VPQR$ is a pyramid on a triangular base in which $PQ = QR = RP = 18$ cm. M is the mid-point of PQ, O the point in MR such that $OM = \frac{1}{3}MR$ and V is vertically above O. If the slant edge $VR = 18$ cm, find

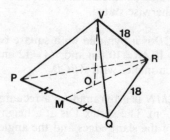

    a) MR *b)* VO *c)* the inclination of VR to the vertical   *d)* VM
    e) the angle between VR and VPQ.
    Give angles to the nearest degree.

110

*9 A rectangular picture frame 30 cm by 50 cm hangs with the long edge BC horizontal and resting against a vertical wall. It is supported by a string attached at two points E and F in the top edge AB, and looped over a nail in the wall at O. G and H are the mid-points of AD and BC. The picture frame makes an angle of $10°$ with the wall. If $EG = GF = 18$ cm and $OE = OF = 30$ cm, find

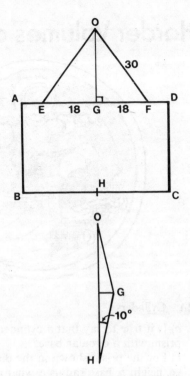

   a) the length of OG,
   b) the distance of the bottom edge BC below O (*Hint* What is angle OHG? Draw triangle OGH to scale.),
   c) the distance of the top edge AD from the wall (to 2 s.f.),
   d) the angle the string OE makes with the wall. Give your answer to the nearest degree. (*Hint* If OX is the projection of OE on the wall, what is the length of EX? What is angle EOX?)

*10 If the string is now attached to points E' and F' 6 cm below E and F, the same nail and the same piece of string still being used, and the picture now hangs at an angle of $12°$ to the wall, calculate

   a) the new distance of the edge BC below O (*Hint* Draw triangle $OG'H$, where G' is the mid-point of $E'F'$. What kind of triangle is $OG'H$?),
   b) the new distance of the edge AD from the wall.

12 Harder Volumes and Surface Areas

12A Cylinders

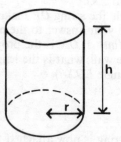

1 *a*) Is it true to say that a cylinder is a prism with a circular base?
b) For the prism shown in the diagram, i.e. height h, base radius r, what is the area of the base? What is the volume?

2 You are given the height and the radius of the base of ten cylinders. Find the volume of each, giving your answer to 3 s.f. and taking π as $3\frac{1}{7}$ or 3·142, whichever you prefer (they give the same answer to 3 s.f.).

   a) 6 cm, 5 cm   *b*) 7 cm, 11 cm   *c*) 11 cm, 7 cm
   d) 2·35 cm, 4·62 cm   *e*) 1 m, 3 m   *f*) 3 m, 1 m

For *g* to *j* give your answer in litres.

   g) 20 cm, 33 cm   *h*) 43 cm, 25 cm   *i*) 2 m, 3 m   *j*) 5 m, $2\frac{1}{2}$ m

3 *a*) A cylindrical tank 2 metres high has a capacity of 1500 litres. Taking π as 3, find the radius of the base.
b) Repeat *a* taking π as 3·142.

4 *a*) A tin of dried milk has a capacity of 1·064 litres. If it is 16 cm high, what is the radius of its base?
b) A tin of evaporated milk has a base of diameter 7·5 cm and holds 442 cm³. What is its height?
c) A jar of instant coffee has an inside radius of 3·3 cm and holds 383 cm³. What is its height?
d) A tin of baked beans has an inside height of 5·5 cm and a capacity of 212 cm³. What is its internal radius?

✱ *5* *a*) If you bought a 2·5 litre tin of paint and found that its height (measured between the lid and the base) was 15·5 cm and its diameter was 15 cm, on opening the tin would you expect to find it very full, nearly full or only three-quarters full?
b) Give a more precise answer to question *a* by saying how many cm below the lid the top of the paint should be.

6 The cylinder in the diagram has a height of 5 cm and the circumference of its base is 12 cm. Imagine a rectangular sheet of thin paper wrapped tightly round the curved surface of the cylinder, so that the two ends just meet and the curved surface is just covered completely.

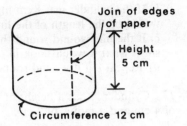

a) What would be the dimensions of this rectangle?
b) What would its area be?
c) What is the area of the curved surface of the cylinder?

7 Repeat question 6, a and c, for each of the following cylinders:

a) Height 6 cm, circumference of base 6 cm.
b) Height 6 cm, radius of base 2 cm (take π as $3\frac{1}{7}$).
c) Height 9 cm, diameter of base 5 cm (take π as 3·142).
d) Height h cm, radius of base r cm.

8 If you answered question 7d correctly, you now know that the area of the curved surface of a cylinder of height h and radius r is $2\pi rh$.

a) What is the area of the top face?
b) What is the area of the bottom face?
c) What is the total area?
d) The expression in c can be factorised. What are the factors?
✳ e) Can you see any advantage in using this factorised form?

9 Find the total surface area of the cylinders whose height and radius are given. Give your answers to 3 s.f.

a) 6 cm, 10 cm   b) 4 cm, 1 cm   c) 11·5 cm, 6·3 cm
d) 14 cm, 2·8 cm   e) 2·8 cm, 14 cm

✳**10** Find the volume and the total surface area of the following cylinders correct to 3 s.f. Ignore any trivial areas. You are given the height (or 'length') and the radius. (f, g and h are hollow but treat them as solid).

a) 1 m, 1 cm   b) 2 mm, 1·25 cm   c) 0·125 mm, 3·8 cm
d) 1 km, 4 cm   e) 35 cm, 2 mm (the ends are cut off)
f) 1·8 m, 7 mm   g) 2·3 m, 6 cm   h) 4 m, 0·9 m (volume only)

✳**11** What do you think the articles listed in question 10 actually are, remembering that the dimensions given are only approximate?

✳**12** If it was possible to 'peel off' the actual surface of a cylinder and lay it on a flat surface, we should get a rectangle the same shape and size as the rectangle of paper we used in question 6. This rectangle is called the 'developed surface' of a cylinder. Now answer the following questions:

a) If A is a point on the circumference of the upper face of a cylinder and B is a point on the circumference of the base directly below A, and a line spirals evenly round the cylinder from A to B, what would this line look like on the 'developed surface' of the cylinder? Draw a sketch to illustrate your answer.

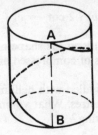

b) If the cylinder was 8 cm high and the circumference of the base was 6 cm, what would be the length of the line? (The cylinder shown is not to scale.)
c) If the line wound round the cylinder twice before reaching *B*, draw a sketch to show what it would look like on the developed surface. What would its length be?
d) Repeat *c* if it wound round 5 times and still finished at *B*.

✱ 13 Cylinder *A* has a height of 7 cm and a radius of 4 cm. Cylinder *B* has a height of 4 cm and a radius of 7 cm.

a) Find the ratio of their volumes.
b) Find the ratio of the areas of their curved surfaces.
c) Find the ratio of their total surface areas.

Note Write your areas and volumes in terms of π. Do not work them out. π will cancel out in the ratios.

✱ 14 A small gasometer has a diameter of 20 m and a height of 20 m when full. Its top surface is curved and its surface area is 10% greater than the area of a flat circle of the same diameter. The gasometer is to be coated outside with a protective coating at a cost of 50p a square metre. What is the total cost of two coats? Answer to the nearest £.

✱ 15 A cylindrical water tank without a lid holds 1·8 m³ of water.

a) If its height is 1·2 metres, what is the radius of its base, correct to 3 s.f.?
b) What is the total outside surface area? What is the total inside surface area?
c) What would be the cost of painting it, inside and outside, using one coat of primer and two coats of paint, the cost of each being 10p a square metre and labour and other costs being ignored?

✱ 16 *a*) If the metal in question 15 is 3 mm thick, what is the approximate volume of metal in the tank?
b) If 1 cm³ of the metal weighs 7 g, allowing 15% extra for rivets and fittings, what is the weight of the empty tank?
c) What is the total weight of tank plus water when the tank is three-quarters full? (One litre of water weighs 1 kg.)

12B Spheres

The volume of a sphere of radius *r* is $\frac{4}{3}\pi r^3$. The area of its surface is $4\pi r^2$.

1 Find the volume and surface area of the following spheres whose radii are given. Where appropriate, give your answer to 3 s.f.

   a) 10 cm   *b*) 7 cm   *c*) 13·5 cm   *d*) 1·2 metres   *e*) 34·8 cm

2 Find the volume and surface area of these spheres whose diameters are given:

   a) 16 cm   *b*) 7·2 cm   *c*) 1 metre   *d*) 4 metres   *e*) 0·14 cm

3 A beach ball is in the shape of a sphere when fully inflated. If its diameter is 50 cm, how many litres of compressed air does it hold?

4 A balloon is filled with helium and it is then in the shape of a sphere with a diameter of 6 metres. What volume of helium does it hold? What is the approximate area of its covering?

5 A hemispherical dog's dish has a diameter of 30 cm. What volume of water would it hold if it was just full?

6 A gas canister is in the form of a cylinder of external diameter 28 cm, with a hemispherical cap at each end. The length of the straight portion is 60 cm. Find

  a) the volume of the straight portion,
  b) the surface area of the straight portion,
  c) the volume and surface area of each cap,
  d) the total volume,
  e) the total surface area.

7 What is the ratio of area to volume for a sphere of radius 6 cm? (Express your volume and area in terms of π without working them out. π will cancel in the final ratio.)

8 Find the ratio of total surface area to volume for:

  a) a cube of side 10 cm,
  b) a cylinder of diameter 10 cm and height 10 cm,
  c) a sphere of diameter 10 cm (remember that π will cancel out).

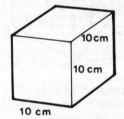
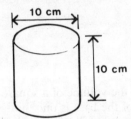
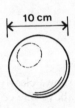

9 For the three solid figures in question 8,

  a) express the volumes as a ratio in the form $p:q:1$,
  b) express the total surface areas as a ratio in the form $p:q:1$.

10 The volume of a heavy sphere is measured by immersing it in water in a graduated jar. It is found to be 256 cm³. Taking π as 3, calculate its radius.

11 Repeat question 10 taking π as 3·14. Give your answer to 3 s.f. (This will involve finding a cube root, either by logs, slide rule or using cube root tables.)

12 The surface area of a tennis ball is 126 cm². What is its radius? What is its volume?

13 A boy has a cricket ball and a larger plastic ball. If the diameters are in the ratio 1:2, what is the ratio   *a)* of their surface areas,   *b)* of their volumes?

✱14 *a)* If in question 13 the ratio of their surface areas is 1:6·25, what is the ratio of their volumes?
  b) If the ratio of their volumes is 1:27, what is the ratio of their surface areas?
  c) If the ratio of their volumes is 1:1·728, what is the ratio of their diameters?

✱15 Three kinds of flavoured chocolate mixtures are being poured into moulds. The moulds are shaped like the three figures in question 8, but the dimensions are 1·5 cm. The mint-flavoured mixture is being poured into cubes, the orange-flavoured mixture into cylinders and the coffee-flavoured mixture into spherical moulds. How many of each kind can be obtained from 1 litre of mix? Which mould is the least practical?

12C Pyramids

1 Construct accurately the net of a square pyramid *VABCD*, in which the base *ABCD* is a square of side 3 cm and the vertex *V* is directly above *A*, so that $AV = 3$ cm. Note that in the net AV_1, AV_2 and the four sides of the square are all 3 cm; also BV_4 and DV_3 are equal to BV_1 and DV_2, both of which are of course equal.

Make three very accurate copies of this net using stiff cartridge paper or thin card, and join them up to give three square pyramids. You will find these can be put together to form a cube of side 3 cm. One way of doing this is shown in the sketch where all the vertices meet at *P* and the pyramids go into the positions *PLMNO*, *PQRNM* and *PSRNO*.

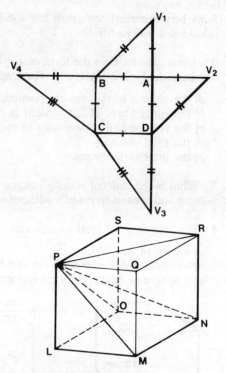

2 Question 1 shows clearly that the volume of a square based pyramid with its vertex directly above one corner of the base is one third of the volume of the cube on the same base. By simple shearing it follows that the volume of any square based pyramid is one third of the cube on the same base. This is an example of a more general rule that says that the volume of *any* pyramid is one third of the area of the base multiplied by the height.

Find the volumes of the following square based pyramids. In each case you are given the length of the side of the base and the perpendicular height.

a) 3 cm, 4 cm   b) 4 cm, 3 cm   c) 5 cm, 6 cm
d) 3·7 cm, 8·1 cm   e) 11 cm, 8·6 cm   f) *a* cm, *b* cm

3 Find the volume of the following pyramids whose bases are rectangles. You are given the lengths of the two sides of the base and the perpendicular height (all in cm).

a) 3, 4, 6   b) 7, 8, 7   c) 9, 11, 10   d) 12, 14, 15   e) 5, 4, 8

4 If the volume of a square based pyramid is 144 cm³ and its height is 12 cm, what is the length of the side of the square base?

5 Repeat question 4 if the volume is 300 cm³ and the height is 9 cm.

6 Repeat question 4 for the following pyramids giving your answers to 3 s.f.

a) Volume 140 cm³, height 7 cm   b) Volume 200 cm³, height 4·7 cm

7 The volume of a pyramid is 120 cm³ and its height is 5 cm. If the base is a rectangle whose length is double its breadth, what are its dimensions?

8 The base of a pyramid is a rectangle 7 cm by 4 cm and its volume is 100 cm³. What is its height, correct to 3 s.f.?

9 A square based pyramid *VABCD* has its vertex *V* directly above *O*, the centre of the base. *AB* is 6 cm and the height *VO* is 5 cm.

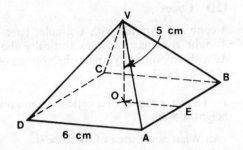

a) If *E* is the mid point of *AB*, what is the length of *OE*?
b) From triangle *VOE*, what is the length of *VE*?
c) What is the area of triangle *VAB*?
d) What is the total area of the five faces of the pyramid?

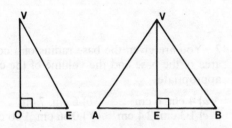

10 Repeat question 9 for a pyramid of height 8 cm on a square base of side 8 cm.

11 A pyramid has a square base *ABCD* of side 4 cm and the vertex *V* is vertically above *A*, *VA* being 3 cm. Find the total surface area of the pyramid. (This time it will help you to draw the net. Compare question 1.)

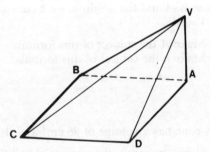

12 Repeat question 11 for a pyramid with *AB*=5 cm and *VA*=12 cm.

13 A pyramid of height 6 cm has a base which is a regular hexagon *ABCDEF* of side 4 cm.

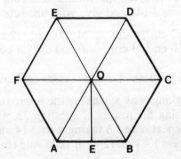

a) What is angle *OAB*? What is the length of *OA*?
b) If *E* is the mid-point of *AB*, what is the length of *OE*? (Use trigonometry or Pythagoras.)
c) What is the area of triangle *OAB*?
d) What is the area of the hexagon?
e) What is the volume of the pyramid?

14 Repeat question 13 with a pyramid of height 9 cm whose base is a regular hexagon of side 6 cm.

15 Could you have deduced the answers to question 14 without carrying out the full calculation? If your answer is *yes* explain how.

12D Cones

A cone is a pyramid with a circular base.
A 'right' cone has its vertex vertically above the centre of the horizontal base.
As with any other pyramid, the volume of a cone is one third of the product of area of base × height.

1 The base radius of a cone is 5 cm and its height is 8 cm.

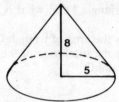

a) What is the area of the base?
b) What is the volume of the cone?

2 You are given the base radius of a cone and its height. In each case state the area of the base and the volume of the cone. Give your answers to 3 s.f. where appropriate.

a) 4 cm, 6 cm   b) 6 cm, 7 cm   c) 7 cm, 6 cm   d) 11·2 cm, 16 cm
e) 1·3 cm, 2·4 cm   f) 0·66 cm, 1·00 cm   g) 27 cm, 55 cm   h) r cm, h cm

3 If the volume of a cone is V, the area of its base is A and the height h, we have seen that $V = \frac{1}{3}Ah$.

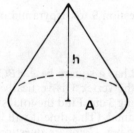

a) Make A the subject of this formula.
b) Make h the subject of this formula.

4 A cone has a volume of 96 cm³ and its height is 6 cm.

a) Using the result of question 3a find the area of the base.
b) What is the radius of the base? Take π as 3.

5 You are given the volume of a cone and its height. Find the radius of the base. Take π as 3·142 and give your answers to 3 s.f.

a) 250 cm³, 5 cm   b) 300 cm³, 12 cm
c) 350 cm³, 3 cm   d) 500 cm³, 8 cm

6 The volume of a cone is 200 cm³ and the radius of the base is 5 cm.

a) Taking π as 3, what is the approximate area of the base?
b) Using the results of question 3b, what is the approximate height of the cone?
c) Repeat a and b taking π as 3·14 and using logarithms or a slide rule. Give your answers to 3 s.f. (2 s.f. if slide rule is used).

7 You are given the volume of a cone and the diameter of the base. Calculate its approximate height, taking π as 3.

a) 240 cm³, 10 cm   b) 60 cm³, 4 cm   c) 900 cm³, 12 cm

d) 100 cm³, 6 cm. This time take π as 3·14, use logarithms or a slide rule and find the height to 3 s.f. (2 s.f. for slide rule).

8 If V is the vertex of a right cone, O the centre of the base and AB any diameter, the angle AVO is known as the 'semi-vertical angle' of the cone. AV is the 'slant height'. If the height of a cone is 6 cm and the semi-vertical angle 30°,

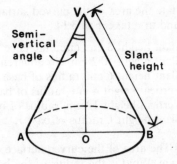

a) what is the radius of the base?
b) what is the approximate area of the base (taking π as 3)?
c) what is the approximate volume of the cone?

9 You are given the height of a cone and its semi-vertical angle. In each case find the radius of the base and the volume. In a, b and c take π as 3 and in d and e take π as 3·14.

a) 6 cm, 20°   b) 8 cm, 15°   c) 11 cm, 33°   d) 1 cm, 45°   e) 4 cm, 16°

10 You are given the radius of the base and the height of a cone. Find the slant height.

a) 3 cm, 4 cm   b) 5 cm, 12 cm   c) 7 cm, 24 cm   d) 10 cm, 12 cm
e) 6 cm, 6 cm

✱ 11 Find the semi-vertical angle (to the nearest degree) and the slant height of the following cones. Take π as 3.

a) Volume $100\,cm^3$, radius of base 5 cm,
b) Volume $200\,cm^3$, radius of base 6 cm,
c) Volume $150\,cm^3$, height 12 cm.

The Surface Area of a Cone

12 Wrap a piece of paper round the curved surface of a cone. Trim the two edges so that they just meet, and trim the paper to fit the bottom edge of the cone. Unwrap the paper and lay it flat on a table. You should have the sector of a circle. (This is the 'developed surface' of a cone. Compare section A question 12.)

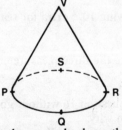
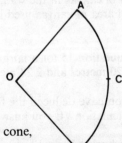

If you now put the paper back on the surface of the cone,

a) what can you say about A and B?
b) what can you say about O and V?
c) what can you say about the slant height of the cone (VP) and the radius of the sector (OA)?
d) what can you say about the circumference of the base of the cone ($PQRS$) and the arc of the sector (ACB)?
e) We saw in chapter 9 that the area of a sector was 'half the length of the arc × the radius'. So what is the area of the curved surface of a cone? (Give your answer in terms of dimensions of the cone, not of the sector.)

13 State the area of the curved surfaces of the following cones. In *c* and *d* take π as 3, and in *e* take π as 3·14.

    a) Slant height 8 cm, circumference of base 12 cm.
    b)·Slant height 10·2 cm, circumference of base 18 cm.
    c) Slant height 6 cm, radius of base 2 cm.
    d) Vertical height 4 cm, radius of base 3 cm.
    e) Vertical height 12 cm, radius of base 5 cm.
    f) Slant height *l*, radius of base *r*.

✴14 *a*) The area of the curved surface of a cone is 20 cm² and the slant height is 4 cm. What is the radius of the base?
    b) The area of the curved surface of a cone is 40 cm² and the radius of the base is 2 cm. What is the slant height?

✴15 A bell tent has a height of 2·75 metres to the top of the pole. The cone of canvas stretches to within 30 cm of the ground. A wall of canvas sewn inside the cone is cylindrical in shape and has a height of 57 cm. The semi-vertical angle of the cone is 40°.

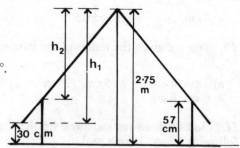

Calculate

 a) the height h_1 and hence the radius of the circle formed by the lower end of the canvas,
 b) the circumference of this circle,
 c) the slant height of the cone of canvas,
 d) the area of the curved surface of the cone of canvas,
 e) the height h_2 and hence the radius of the cylindrical wall of canvas,
 f) the circumference of this wall,
 g) the area of canvas in the wall,
 h) the total area of canvas used in the tent, allowing 10% extra for seams and overlaps.

✴16 Repeat question 15 for a fairground tent where the three measurements given are 11 metres, 1·20 metres and 2·28 metres.

✴17 Could you have deduced the final answer to question 16 without going through the whole calculation? If you answer is *yes*, explain why.

✴18 A plastic bucket has a height of 24 cm, the radius of the top of the bucket is 12 cm and the radius of the bottom is 9 cm. The bucket could be considered as the section of a cone *VABCD* whose top had been cut off. The top portion *VEFGH* is also a cone.

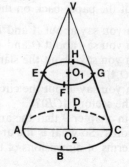

 a) If the height of the smaller cone VO_1 is *h*, what is the height VO_2 of the larger cone in terms of *h*?

120

b) By comparing triangles VO_1G and VO_2C find an equation for h and hence find the value of h.

c) Using this value, what is the volume of the smaller cone? (Take π as 3.)

d) What is the volume of the larger cone?

e) What is the volume of the bucket?

f) The section of a cone cut off by two parallel planes is known as a 'frustrum' of a cone. Is the bucket a frustrum?

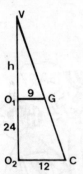

*19 Repeat question 18 for a child's bucket of height 12 cm and top and bottom diameters 12 cm and 8 cm respectively.

20 A funnel is in the form of an inverted cone with a semi-vertical angle of 28°. The stem contains a tap immediately below the bottom of the funnel.

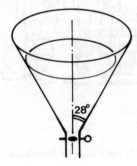

a) When the water is 4 cm deep and the tap is closed, what is the radius of the circle formed by the top surface of the water?

b) What is the area of this circle?

c) What is the volume of the water in the funnel?

*21 Repeat question 20 for a depth of water of

a) 3 cm   b) 5 cm   c) 6 cm

*22 Repeat question 20 if the semi-vertical angle is 20°.

*23 Sherry glasses in the form of inverted cones with semi-vertical angles of 24° are being filled to a depth of 7 cm.

a) How much sherry will each glass hold?

b) How many glasses can be filled from a bottle containing one litre of sherry?

13 Algebra

13A Multiplying Pairs of Brackets

Multiply the brackets and simplify the following expressions:

1 $x(x+1)+2(x+1)$ 2 $x(x+1)-2(x+1)$ 3 $x(x-3)+2(x-3)$

4 $x(x-3)-2(x-3)$ 5 $x(x+4)+7(x+4)$ 6 $x(x+7)+4(x+7)$

7 $x(x-2)+5(x-2)$ 8 $x(x+5)-2(x+5)$ 9 $x(x-3)-6(x-3)$

10 $x(x-6)-3(x-6)$ 11 $x(x-7)+2(x-7)$ 12 $x(x-7)-2(x-7)$

13B

By splitting the first brackets and multiplying the second bracket by each term in the first, find the following products:

1 $(x+2)(x+3)$ 2 $(x+1)(x+4)$ 3 $(x+2)(x+7)$

4 $(x+5)(x+3)$ 5 $(x+6)(x+6)$ 6 $(x-2)(x-4)$

7 $(x-1)(x-5)$ 8 $(x-3)(x-9)$ 9 $(x-8)(x-3)$

10 $(x-3)(x-3)$ 11 $(x+6)(x-5)$ 12 $(x+4)(x-2)$

13 $(x+2)(x-8)$ 14 $(x+3)(x-9)$ 15 $(x-6)(x+2)$

16 $(x-4)(x+3)$ 17 $(x-7)(x+8)$ 18 $(x-2)(x+7)$

19 $(x+2)(x-2)$ 20 $(x-5)(x+5)$

13C

1 The diagram shows a rectangle which measures $(x+5)$ units by $(x+4)$ units. The area of the rectangle is divided up to show $(x+5)(x+4) = x^2 + 5x + 4x + 20$, i.e. $(x+5)(x+4) = x^2 + 9x + 20$.
Repeat this with a diagram to show the area of a rectangle which measures $(x+3)$ units by $(x+1)$ units.

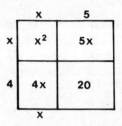

2 Draw similar diagrams to show   *a)* $(x+4)^2$ *b)* $(x+3)(x+2)$

✱*3* Can you think of diagrams to show these multiplications? They require more ingenuity.

   a) $(x+2)(x-1)$ *b)* $(x-2)(x-3)$ *c)* $(x-3)^2$ *d)* $(x+2)(x-2)$

✱*4* Draw diagrams to show

   a) $(x+a)(x+b)$ *b)* $(x+m)(x-n)$ *c)* $(x+y)(x-y)$

13D

Find the products of the following:

1 $(x+5)(x+1)$ *2* $(x+2)(x+4)$ *3* $(x-2)(x-3)$

4 $(x-6)(x-2)$ *5* $(x+2)(x+2)$ *6* $(x-7)(x-7)$

7 $(x+2)(x-1)$ *8* $(x+5)(x-4)$ *9* $(x-7)(x+6)$

10 $(x-6)(x+7)$ *11* $(x+7)(x+1)$ *12* $(x+3)(x+8)$

13 $(x-8)(x+4)$ *14* $(x+2)(x-9)$ *15* $(x-9)^2$

16 $(x-8)(x+8)$ *17* $(x+12)(x+1)$ *18* $(x+5)^2$

19 $(x+10)(x-3)$ *20* $(x-10)(x-2)$

13E

Find the products of the following:

1 $(x+2)(x+5)$ *2* $(x+1)(x+7)$ *3* $(x-3)(x-5)$

4 $(x-5)(x-8)$ *5* $(x+8)(x-5)$ *6* $(x-8)(x+5)$

7 $(x+3)(x+7)$ *8* $(x-3)(x-7)$ *9* $(x-3)(x+7)$

10 $(x+3)(x-7)$ *11* $(x+6)(x-8)$ *12* $(x+4)(x-4)$

13 $(x-1)^2$ *14* $(x+7)(x-4)$ *15* $(x+3)(x+9)$

16 $(x+8)(x-1)$ **17** $(x-2)(x-5)$ **18** $(x-3)(x+4)$

19 $(x-4)(x+3)$ **20** $(x-1)(x+12)$ **21** $(x-3)(x+3)$

22 $(x+5)(x+7)$ **23** $(x-5)(x-7)$ **24** $(x+4)(x-9)$

25 $(x-2)(x-18)$ **26** $(x-3)(x-12)$ **27** $(x+3)(x-12)$

28 $(x-6)^2$ **29** $(x-14)(x+2)$ **30** $(x-2)(x+11)$

13F

Find the products of the following:

1 $(x+y)(x+2y)$ **2** $(x-2a)(x-5a)$ **3** $(a+5b)(a+b)$

4 $(l-m)(l-2m)$ **5** $(p-3q)(p+7q)$ **6** $(a+8b)(a-b)$

7 $(s-5t)(s+6t)$ **8** $(x-7y)(x-y)$ **9** $(x+2y)(x-2y)$

10 $(l-6m)(l+6m)$ **11** $(2x+3)(x+1)$ **12** $(3x-1)(2x-1)$

13 $(3x-5)(2x+3)$ **14** $(5x+1)(3x+1)$ **15** $(4x+1)(3x-1)$

16 $(4x+3)(2x+1)$ **17** $(3x-2)(3x+2)$ **18** $(6x-1)(x+2)$

19 $(2x-5)(3x+2)$ **20** $(4x+3)(2x+3)$

13G

Find the products of the following:

1 $(x+3y)(x-5y)$ **2** $(3a+1)(2a-3)$ **3** $(a+9)(a-5)$

4 $(2-b)(3+b)$ **5** $(2l+5m)(l-3m)$ **6** $(10+x)(2-x)$

7 $(1-7y)^2$ **8** $(p+3q)(5p-2q)$ **9** $(1-a)(1+a)$

10 $(3-x)(5-2x)$ **11** $(l-m)(3l-m)$ **12** $(2p+1)(1-p)$

13 $(4-3x)(1+x)$ **14** $(6x+1)(2x-3)$ **15** $(5x-3)(2x+3)$

13H

Complete the following:

1 $(x+\quad)^2=x^2+\quad+9$ **2** $(x-\quad)^2=x^2-2x+$

3 $(x-\quad)^2=x^2-\quad+25$ **4** $(x+\quad)^2=x^2\quad+49$

5 $(2x\quad)^2=4x^2-\quad+9$ **6** $(2x\quad)^2=4x^2-4x$

124

7 $(x \quad)^2 = x^2 + 12x$

8 $(x \quad)^2 = x^2 - \quad + 100$

9 $(x \quad)^2 = x^2 + x$

10 $(x+2)(x \quad) = x^2 \quad + 10$

11 $(x+3)(x \quad) = x^2 + 7x +$

12 $(x+3)(x \quad) = x^2 \quad - 15$

13I

Multiply out the following:

1 $2(x^2 - 5)$

2 $3x(x+1)$

3 $2a(3a - 2b + c)$

4 $5ab(1 - 2a + 4b)$

5 $a^2b(a + 3ab - 5b)$

6 $x^3(1 + 2x - 4x^2 + x^3)$

7 $3ab(a - 2b)$

8 $a(a+b)(a+2b)$

9 $x(x+5)(x-4)$

10 $ab(1+a)(1+b)$

11 $x(1+x)(3-x)$

12 $a(a+2b)(a-5b)$

13J Factors

Factorise the following by finding common factors:

1 $2a^2b + 3ab^2$

2 $4a^2 - 8ab$

3 $3a^3 - 2a^2b + 6ab^2$

4 $a^4 - 3a^3 + a^2$

5 $a^2 - 4a^3 - 7a^4$

6 $a^3b^2 - 3a^2b^3 - 5ab^4$

7 $2x^2 - 6x$

8 $5x^2 + 10$

9 $12x + 4x^2$

10 $5x^5 + 4x^4 - 6x^3 + x^2$

11 $3x^2 + 2x^3 - 5x^4 + x^5$

12 $4x^2y^4 + 2xy^5 - 6x^3y^3$

13K

Factorise the following expressions by writing each as the product of a pair of brackets:

1 $x^2 + 3x + 2$

2 $x^2 + 5x + 6$

3 $x^2 + 5x + 4$

4 $x^2 + 7x + 12$

5 $x^2 + 7x + 10$

6 $x^2 + 7x + 6$

7 $x^2 - 2x + 1$

8 $x^2 - 4x + 4$

9 $x^2 - 8x + 12$

10 $x^2 - 8x + 15$

11 $x^2 - 7x + 10$

12 $x^2 - 7x + 6$

13 $x^2 + x - 2$

14 $x^2 + 2x - 15$

15 $x^2 - 2x - 15$

16 $x^2 + 6x - 7$

17 $x^2 - 6x - 16$

18 $x^2 - 16$

19 $x^2 + 15x - 16$

20 $x^2 - 3x - 10$

13L

Factorise the following expressions:

1 $x^2 + 4x$

2 $x^2 - 4$

3 $x^2 + 4x + 3$

4 x^2-9	**5** x^2-9x	**6** $x^2-9x+18$
7 x^2+5x-6	**8** x^2-5x+6	**9** x^2+6x+8
10 x^2-6x+8	**11** x^2-2x-8	**12** $x^2-2x-24$
13 $x^2+2x-35$	**14** $x^2+12x+35$	**15** $x^2-10x-24$
16 $x^2-14x+24$	**17** $x^2-14x+48$	**18** $x^2-14x-15$
19 $x^2+14x+40$	**20** $x^2+13x+40$	**21** $x^2+5x-14$
22 $x^2-5x-24$	**23** $x^2-13x+22$	**24** x^2-49
25 $x^2-12x+36$	**26** x^2-121	**27** $x^2-16x+64$
28 x^2-8x	**29** $x^2-12x+32$	**30** $x^2-14x-32$

13M Harder Factors

Factorise the following expressions:

1 $x^2+2xy+y^2$	**2** $a^2+12ab+11b^2$	**3** p^2-q^2
4 $l^2-9lm+20m^2$	**5** r^2-16s^2	**6** $p^2-4pq-21q^2$
7 $a^2b^2+3ab-18$	**8** $6+x-x^2$	**9** $20-9x+x^2$
10 $10-3x-x^2$	**11** $12+4x-x^2$	**12** $25-4a^2$
13 $3x^2-12$	**14** $5a^2-5$	**15** $12b^2-3$
16 $3x^2+6x+3$	**17** $5x^2+20x+20$	**18** $2x^2+6x-20$
19 $2x^2-10x-12$	**20** $4x^2-4x-8$	**21** $6x^2+5x+1$
22 $6x^2+11x+3$	**23** $6x^2+19x+3$	**24** $10x^2+3x-1$
25 $5x^2-11x+2$	**26** $5x^2+3x-2$	**27** $4x^2+12x+9$
28 $8x^2+2x-3$	**29** $8x^2+10x-3$	**30** $4x^2-x-3$

13N

Using the factors of the difference of two squares, find the value of each of the following:

1 85^2-15^2	**2** 97^2-3^2	**3** 112^2-12^2	**4** 35^2-15^2
5 $15 \cdot 5^2-4 \cdot 5^2$	**6** 992^2-8^2	**7** $0 \cdot 88^2-0 \cdot 12^2$	**8** $7 \cdot 8^2-2 \cdot 2^2$
9 $0 \cdot 65^2-0 \cdot 15^2$	**10** $2 \cdot 7^2-0 \cdot 7^2$		

4 Fractions in Algebra

14A

Fill in the blank spaces in each of the following to make the fractions equivalent.

1 $\dfrac{x}{3}=\dfrac{2x}{}=\dfrac{x}{15}$

2 $\dfrac{3x}{7}=\dfrac{6x}{}=\dfrac{x}{42}$

3 $\dfrac{5x}{2y}=\dfrac{}{6y}=\dfrac{20}{}$

4 $\dfrac{6}{x}=\dfrac{18}{}=\dfrac{}{5x^2}$

5 $\dfrac{2}{9x^2}=\dfrac{2x}{}=\dfrac{6y}{}$

6 $\dfrac{7x}{3y}=\dfrac{}{3y^2}=\dfrac{14x^3}{}$

7 $\dfrac{x-1}{3}=\dfrac{2(x-1)}{}$

8 $\dfrac{2(x+1)}{x}=\dfrac{}{x(x+4)}$

9 $\dfrac{2x+3}{x}=\dfrac{}{3x^2}$

10 $\dfrac{3(x-2)}{4}=\dfrac{x(3x-6)}{}$

11 $\dfrac{5}{3(x+1)}=\dfrac{5(x+2)}{}$

12 $\dfrac{6(x+1)}{x^2-1}=\dfrac{}{x-1}$

14B

Solve the following equations:

1 $\dfrac{x}{3}+\dfrac{x}{5}=4$ (*Hint*   Multiply every term in the equation by 15.)

2 $\dfrac{5x}{12}-\dfrac{x}{3}=1$ (*Hint*   Multiply every term in the equation by 12.)

3 $5-\dfrac{3x}{5}=x$

4 $\dfrac{4x}{7}-\dfrac{1}{2}=\dfrac{3x}{4}$

5 $2(\dfrac{x}{5}-1)=\dfrac{3x}{8}$

6 $\dfrac{11}{12}+\dfrac{7x}{8}=\dfrac{5x}{6}+\dfrac{3}{4}$

7 $\dfrac{2x}{5}-\dfrac{3}{4}=\dfrac{1}{10}+\dfrac{3x}{8}$

8 $\dfrac{7x}{12}=3(1-\dfrac{x}{4})$

9 $\frac{2}{3}(x+2)=\frac{1}{4}(3x+5)$ **10** $\frac{3}{8}(1-4x)=2$ **11** $\frac{5}{9}(x-6)=\frac{2}{3}$

12 $\frac{1}{5}(2x-3)=\frac{1}{3}(4x-1)$ **13** $4-\frac{2}{7}(3x+5)=0$ **14** $\frac{7x}{12}-\frac{3}{4}(x-1)=0$

15 $1\frac{1}{6}=\frac{4}{9}x+\frac{2}{3}$ **16** $2\frac{1}{4}-\frac{3}{8}(x-1)=0$ **17** $\frac{x+1}{2}=\frac{x-5}{3}$

18 $\frac{2x-3}{4}=\frac{x-7}{6}$ **19** $\frac{3-5x}{9}=\frac{3-2x}{6}$ **20** $\frac{4+3x}{7}=1\frac{2}{3}$

14C

Solve the following equations:

1 $\frac{2x}{5}-\frac{3}{20}=\frac{7x}{10}$ **2** $\frac{1}{5}(2x-7)=\frac{4}{15}$ **3** $\frac{1}{9}(4-5x)=2\frac{1}{6}$

4 $\frac{3-8x}{12}=2\frac{1}{4}$ **5** $\frac{2x-1}{6}=\frac{8+3x}{15}$ **6** $4-\frac{1}{5}(x-3)=0$

7 $3+\frac{x-7}{9}=0$ **8** $\frac{5-2x}{8}=\frac{7+x}{12}$ **9** $\frac{6x+5}{2}+\frac{x+5}{3}=0$

10 $\frac{8x-1}{3}-\frac{4x+11}{9}=0$ **11** $\frac{5x+1}{6}+\frac{2x+1}{3}=1$ **12** $\frac{4-x}{8}-\frac{1-x}{4}=3$

13 $\frac{2x-1}{3}-\frac{3x-5}{4}=1$ **14** $\frac{4x-1}{5}+1\frac{1}{2}=\frac{3x}{10}$ **15** $x-\frac{x+1}{4}=\frac{x-2}{3}$

16 $\frac{x+1}{6}-\frac{x-5}{3}=2$ **17** $\frac{4+3x}{5}-\frac{2-x}{4}=x$ **18** $\frac{3x-1}{4}=1-\frac{x+2}{6}$

19 $4-\frac{3-2x}{15}=\frac{5+2x}{3}$ **20** $\frac{x}{9}-\frac{4x-3}{12}=\frac{5x}{18}$

✳ 14D

Solve the following equations:

1 $\frac{2}{5}(x+2)=4$ **2** $\frac{2}{7}(2x+5)=1$

3 $2-\frac{5}{8}(3-2x)=0$ **4** $\frac{5}{6}=\frac{2}{9}(1+2x)$

5 $1\frac{3}{4}-\frac{5}{12}(1-4x)=0$ **6** $\frac{5}{9}x-\frac{2}{3}(1+x)=1\frac{1}{2}$

7 $\dfrac{2(2x-1)}{5}=\dfrac{3(1-2x)}{7}$

8 $\dfrac{5}{9}(x-3)-\dfrac{2}{3}(x-4)=0$

9 $\dfrac{7}{8}(1-2x)-\dfrac{3}{4}(2-3x)=1$

10 $5-\dfrac{3}{10}(3x-2)=\dfrac{2}{5}(2x-3)$

11 $\dfrac{7}{12}(x+4)-\dfrac{3}{8}(x+2)=3$

12 $\dfrac{2(3x-1)}{7}-\dfrac{3(x-1)}{4}=1$

13 $5-\dfrac{2(x-5)}{3}=\dfrac{3x+1}{6}$

14 $\dfrac{5(5-2x)}{6}-1\dfrac{3}{5}=\dfrac{7(1-x)}{15}$

15 $\dfrac{7}{8}x+\dfrac{5}{12}(1-3x)=1\dfrac{1}{6}$

16 $\dfrac{3(4x-3)}{10}-\dfrac{2}{3}x=\dfrac{2(7x-3)}{15}$

*17 $\left.\begin{array}{l}\dfrac{x}{3}+\dfrac{y}{4}=4\\[2mm]\dfrac{x}{2}+\dfrac{y}{3}=5\dfrac{2}{3}\end{array}\right\}$

*18 $\left.\begin{array}{l}\dfrac{3x}{4}+\dfrac{y}{2}=6\dfrac{1}{2}\\[2mm]\dfrac{x}{3}+\dfrac{y}{4}=3\end{array}\right\}$

*19 $\left.\begin{array}{l}\dfrac{2x}{5}-\dfrac{2y-3}{10}=3\\[2mm]x-y=3\dfrac{1}{2}\end{array}\right\}$

*20 $\left.\begin{array}{l}\dfrac{x-3}{2}+\dfrac{y+3}{3}=2\\[2mm]\dfrac{2x}{5}-\dfrac{y+6}{6}=1\end{array}\right\}$

14E

Write each of the following as a fraction in its simplest form:

1 $\dfrac{x}{2}+\dfrac{3x}{5}$

2 $\dfrac{2x}{9}+\dfrac{x}{3}$

3 $\dfrac{3x}{14}+\dfrac{2x}{7}$

4 $\dfrac{2x}{5}-\dfrac{3x}{10}$

5 $\dfrac{7x}{8}-\dfrac{5x}{6}$

6 $\dfrac{9x}{14}-\dfrac{3x}{7}$

7 $\dfrac{7x}{18}+\dfrac{x}{2}+\dfrac{5x}{6}$

8 $\dfrac{3x}{4}-\dfrac{x}{8}+\dfrac{5x}{6}$

9 $\dfrac{x+1}{2}+\dfrac{x+2}{3}$

10 $\dfrac{x}{2}+\dfrac{2x-1}{4}$

11 $\dfrac{2x-3}{5}+\dfrac{x-1}{4}$

12 $\dfrac{3x-1}{10}-\dfrac{2x+1}{15}$

13 $\dfrac{3x+5}{8}-\dfrac{5x+1}{6}$

14 $\dfrac{1+3x}{8}+\dfrac{1-2x}{16}$

15 $1\dfrac{1}{2}-\dfrac{x+3}{5}$

16 $\dfrac{2x-5}{9}+\dfrac{3x+2}{6}$

17 $\dfrac{5x-7}{12}-\dfrac{3x+2}{8}$

18 $\dfrac{2x-1}{5}-\dfrac{3x-2}{6}$

19 $\dfrac{9+2x}{12}+\dfrac{1+2x}{4}$

20 $\dfrac{5+2x}{9}-\dfrac{3-4x}{12}$

14F

Simplify the following:

1 $\dfrac{2x}{3}-\dfrac{x+1}{5}$

2 $\dfrac{2(x+2)}{3}-\dfrac{3(x+1)}{5}$

3 $\dfrac{3(x-3)}{7}+\dfrac{2(x+2)}{3}$

4 $\dfrac{5(1-x)}{9}+\dfrac{5(1+x)}{6}$

5 $2\dfrac{2}{3}-\dfrac{7}{12}(x+2)$

6 $\dfrac{3}{4}(2x-1)-\dfrac{7}{10}(3x-2)$

7 $\dfrac{2(3x-1)}{7}-\dfrac{(3x-2)}{2}$

8 $\dfrac{3(2x+3)}{4}-\dfrac{5(2x-1)}{8}$

9 $\dfrac{5x}{6}+\dfrac{3(2-x)}{8}$

10 $\dfrac{3}{10}(5x-1)+\dfrac{1}{6}(3x+1)$

11 $1\dfrac{1}{2}-\dfrac{4}{9}(3-x)+\dfrac{7x}{18}$

12 $\dfrac{5}{6}-\dfrac{1}{15}(3x-1)-\dfrac{7x}{10}$

14G Fractional Equations

1 I think of a number, add 2 and divide the result by 4. This gives one third of the original number. What was it?

2 I think of a number, subtract 3 and multiply the result by 2. I find that this gives the same result as adding 3 to the original number and then dividing by 2. What number did I first think of?

3 If I buy 4 bars of chocolate at $11\frac{1}{2}$p each and 5 at x pence each, the average cost per bar is 9p. What is x?

4 Susan saves $\frac{1}{4}$ of her pocket money each week, and spends $\frac{1}{2}$ of the remainder on magazines and 6p on sweets. She still has 24p left. How much pocket money does she receive?

5 I buy x books at a total of £6 and find that one book costs the same as 8 notebooks at $3x$ pence each. How many books did I buy?

6 Ann is two years older than her sister, Judith. In a game, Ann is asked to double her age and then add 6. She then divides the result by 8. Judith finds that the answer is exactly a third of her own age. How old are they?

7 I think of a number, add 6 and divide the result by 4. This is 5 less than doubling the original number, subtracting 3 and then dividing by 3. What is the number?

8 On a journey a motorist travels 130 km on a motorway at an average speed of 104 km per hour and for another 45 minutes on minor roads. If his average speed for the whole journey is 84 km per hour, how far did he travel on minor roads?

9 A greengrocer sells the first 15 melons out of a crate of 24, and then finds that the rest have deteriorated and must be sold quickly. He sells the remainder at 30p less per melon than he was charging originally. They all sell at this reduced price and in the end he finds that he has averaged $\frac{3}{4}$ of the original price per melon. What was the original price?

10   In a fourth form there are four more girls than boys. Three-eighths of the girls and two-thirds of the boys ride bicycles to school. Half the class ride bicycles to school. How many boys and how many girls are there in the class?

11   Two boys have started collecting stamps. The table shows what fraction of each boy's collection is British, American, etc.

British	$\frac{1}{3}$	$\frac{1}{4}$
Other European	$\frac{1}{4}$	$\frac{2}{5}$
American	$\frac{1}{6}$	$\frac{1}{3}$
Asian	$\frac{1}{12}$	$\frac{1}{60}$
African	$\frac{1}{6}$	NIL

If the number of stamps in the first boy's collection is half the number in the second boy's collection, and if the first boy has 30 more Asian stamps than the second boy, how many stamps are there in each collection?

12   If the first boy has twice as many stamps as the second, and the number of British plus other European stamps he has is 279 more than the corresponding number in the second boy's collection, how many stamps does each have?

13   Two school parties on the Pennine walk start from different hostels in the morning but reach the same hostel at night. One party find that one eighth of their journey that day is a stretch of bog. The other party walk eight kilometres further, but the only bog they meet is the same stretch as the first party. One tenth of their journey is bog. How far does each party walk that day?

14   At a school dance there are eight more girls than boys. Only couples are allowed to take the floor, i.e. boys must dance with girls and girls cannot dance together. During a tango, four-fifths of the boys and three-quarters of the girls take the floor. How many girls are at the dance?

*15   If 15 is subtracted from a certain number of two digits, the number obtained is one third of the number formed by reversing the digits. The tens digit of the first number is half the units digit. What are the numbers? (It may help you to look at 1E question 16.)

*16   A number of two digits is divided by 4 and 45 is added. This gives the number obtained by reversing the digits. If 9 is added to the original number, the number obtained is seven times the sum of the digits. What was the original number?

*17   A sum of money is to be divided into 3 parts so that A receives $\frac{1}{3}$ of the total and B receives twice as much as C. If A had only taken $\frac{1}{4}$ of the total, and B had still taken twice as much as C, then C would have had £1 more than before. How much money was there?

18   A circle is divided into two sectors, one less and one greater than a semicircle. The smaller sector is divided into five equal parts. The larger sector is divided into 8 equal parts, each of which is 6° larger than those in the smaller sector. What was the angle of the original larger sector?

Miscellaneous Examples B

B1

1 A tank contains 1500 litres of water. If the base is a rectangle which measures 5 m by 2 m, find the depth of the water.

2 The diagram shows the end of a prism of length 20 cm. $DC = 4$ cm, and V is 10 cm above BC. $\angle VAD = \angle VDA = 50°$. Calculate the area of the end face and the volume of the prism.

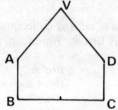

3 Find the gradient of the line joining (6,0) to (2,2). If the line is produced (extended) to cut the y axis at A, find the co-ordinates of A and hence write down the equation of the line.

4 Solve the equations:

a) $\dfrac{1}{2}x + 3 = x - 5$ b) $2(x-1) + 3(x-7) = 2$ c) $\dfrac{x}{4} - \dfrac{(x-2)}{6} = 0$

5 Write without brackets and simplify:

a) $2(x+3) - 3(x-2) + (x+4)$ b) $(x+1)(x-2)$
c) $(2x-3)(3x-2)$ d) $(2x+4)(2x-4)$
e) $x(x^2 + 2x + 3) - 3(x+2)(x-1)$

6 a) Find the area of the trapezium shown in the diagram.

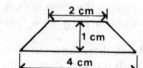

b) The second diagram shows a net for a square based pyramid. All the edges of the pyramid are 4 cm long. The four tabs are each the same as the trapezium above. Find the area of card used and the area of the smallest square of card from which this net can be cut.

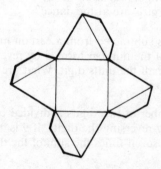

B2

1 What is the gradient of the line $2y = x - 5$? If the line cuts the x and y axes at B and C, respectively, what is the area of triangle OBC? Find the equation of the line parallel to BC which passes through the origin O.

2 Simplify   a) $\dfrac{2}{5} + \dfrac{1}{6}$ b) $\dfrac{2}{a} + \dfrac{1}{b}$ c) $\dfrac{2x+3}{5} - \dfrac{x-1}{6}$

3　Factorise　a) $ax^3 + 5ax^2 + 6a^2x$　　b) $x^2 - 5x + 4$
　　　　　　　c) $3x^2 - 9x - 12$　　d) $2x^2 + 5x - 12$

4　Find the area of a circle of radius 3 cm. Take $\pi = 3\cdot14$, and give your answer correct to 3 s.f.
Which of the following has the smallest area and which has the largest?

a) a semicircle of radius 3 cm
b) a rectangle 5·2 cm by 2·7 cm
c) the triangle shown in the diagram
d) a square of side 3·7 cm

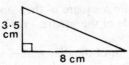

Do the two shapes which you have found also have the shortest and longest perimeters?

5　Use logarithms to find the values of:　　a) $\dfrac{74\cdot8}{5\cdot68}$　　b) $\dfrac{48\cdot3}{9\cdot72 \times 7\cdot54}$　　c) $\sqrt[3]{0\cdot0517}$

6　$ABCD$ is a square of side 12 cm, $LBMN$ is a square of side 8 cm, L is a point on AB and M is on BC. $DPNQ$ is a square of side 4 cm and P is on AD. Calling $ABCD$ square I, $LBMN$ square II and $DPNQ$ square III, give the centres of enlargement and the scale factors for the transformations which

a) map II on to I　　b) map II on to III　　c) map I on to III.

B3

1　What is the gradient of the line joining $A(1,2)$ to $B(4,5)$? If this line is reflected in $y = 0$ to give $A'B'$, find the gradient of $A'B'$.

2　Find the perimeter of a 35° sector cut from a circle of radius 7 cm. Take $\pi = 3\frac{1}{7}$.

3　Solve the equations　a) $\frac{1}{2}x + 6 = 9$　　b) $3x^2 = 48$　　c) $2(x - 5) = 5 + 3(3 - 2x)$

4　Factorise　a) $x^2 + 4x + 4$　　b) $x^2 + 3x - 54$　　c) $x^2 - 64$

5　The diagram shows a pyramid on a triangular base ABC. V the vertex is vertically above A. If $VC = 12$ cm, $AC = 10$ cm, $BC = 6$ cm and $AB = 8$ cm, calculate

a) angle ABC,
b) the height VA,
c) the angle between BV and the base,
d) the angle between the planes VCA and VBA.

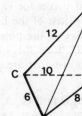

6　A child has two building bricks, one of which is a cube of side 4 cm and the other a prism of height 2 cm and a square cross section of area 36 cm². Sketch a possible shape for a third brick which with the other two could be used to make a cube of side 6 cm. Mark its measurements clearly, and calculate its volume.

B4

1　Write without brackets:　a) $(x - 5)(x + 4)$　　b) $(x + 3)(x - 7)$　　c) $(3x + 1)(x + 3)$

2 What is the gradient of the line $y = 3x + 5$? Write down the equations of lines parallel to this line but passing through

  a) (0,0)   *b)* (0,1)   *c)* (2,0)

3 How many discs of radius 2 cm can be cut from a piece of metal 12 cm by 16 cm? Draw a sketch to show how this can be done. The remaining metal is melted down and reshaped into a square of the same thickness as the original rectangle. Find the length of the side of the square. ($\pi = 3 \cdot 14$.)

4 Find the length of the diagonal of a rectangular box of height 4 cm whose base measures 3 cm by 8 cm. Find also the inclination of the diagonal to the base. Give your answer to the nearest degree.

5 On graph paper draw the x and y axes for values from -3 to 5. Plot the points (0,0), (3,0) and (2,1). Join up and call the triangle S. On the same axes show

  S_1, the image of S after reflection in $x = 0$,
  S_2, the image of S after rotation of $180°$ about 0,
  S_3, the image of S after reflection in $y = x$,
  S_4, the image of S after translation by the vector $\begin{pmatrix} 2 \\ -2 \end{pmatrix}$.

  a) Find the mirror line which reflects S_3 on to S_4.
  b) Describe a transformation which maps S_2 on to S_4.
  c) Describe a transformation which maps S_2 on to S_3.

6 Solve the following equations:

  a) $2(x+3) - 3(1-2x) = 5$ *b)* $\dfrac{x}{3} + 1\tfrac{1}{2} = \dfrac{3x}{4}$

  c) $\dfrac{2(1+x)}{5} - \dfrac{3(4-2x)}{4} = \dfrac{3x-10}{10}$

B5

1 Draw the x and y axes for values from -2 to 5. Plot the points $A(-1,2)$ and $B(2,5)$. Find the gradient of the line AB, the point where it cuts the y axis and hence the equation of AB. C is the point $(2,-1)$. If $x = 2$ is the line of symmetry of the quadrilateral $ABCD$, draw this quadrilateral and write down the co-ordinates of D. Describe $ABCD$ and find its area. Write down the equations of the lines which pass through

  a) A and C *b)* D and C *c)* B and D.

2 The diagram shows the shape of a small warehouse. Calculate

  a) the distance down the slope of the roof,
  b) the area of the roof,
  c) the area of one end,
  d) the volume of the warehouse.

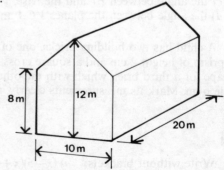

3 Factorise *a)* $4x^3 + 8x$ *b)* $x^2 + 13x + 42$ *c)* $2x^2 + 11x - 6$ *d)* $x^2 - 4$
e) $x^2 + 4x$

4 A cylindrical container 10 cm high holds 250 cm³ of water. A similar container is 20 cm high. How much water does it hold?

5 A pyramid *VABCD* is on a rectangular base *ABCD* in which *AB* = 5 cm and *AD* = 7 cm. The vertex *V* is vertically above the centre of the base and the slant edges are 8 cm long. *E* and *F* are the mid points of *AB* and *AD*. Find the lengths of *VE* and *VF*. Calculate also the angles between *a)* *VA* and the base, *b)* the face *VAB* and the base, *c)* the face *VAD* and the base, *d)* the faces *VAB* and *VCD*. Give your answers to the nearest half degree.

6 Find the radius of the base of a cone of volume 450 cm³ and height 11·75 cm. Give your answer correct to 2 s.f.

15 Quadratic Equations

15A

Solve the following equations, all of which are given in factor form:

1 $x(x-3)=0$ **2** $x(x+1)=0$ **3** $2x(x-5)=0$

4 $4x(2-x)=0$ **5** $(x-1)(x-3)=0$ **6** $(x+2)(x-7)=0$

7 $(x+4)(x-4)=0$ **8** $(x+7)(x+9)=0$ **9** $3(x+2)(x-4)=0$

10 $x(2x-3)=0$ **11** $4x(3x-1)=0$ **12** $5(2x-1)(x-2)=0$

13 $2(4-3x)(1+2x)=0$ **14** $(x+5)^2=0$ **15** $6(1-4x)^2=0$

15B

Solve the following equations by first finding factors. Check your answers to the first ten. Remember there are two answers to every question.

1 $x^2-x=0$ **2** $x^2+8x=0$ **3** $2x^2-8x=0$

4 $9x-3x^2=0$ **5** $x^2-3x+2=0$ **6** $x^2-7x+12=0$

7 $x^2+11x-12=0$ **8** $x^2+11x+10=0$ **9** $x^2+10x-11=0$

10 $x^2-144=0$ **11** $x^2-24x+144=0$ **12** $x^2-24x=0$

13 $x^2+9x+14=0$ **14** $x^2-9x-36=0$ **15** $x^2-36=0$

16 $x^2-10x+21=0$ **17** $x^2+4x-21=0$ **18** $x^2+x-42=0$

19 $4x^2+12x+8=0$ **20** $5x^2+5x-10=0$

15C

Solve the following equations. Remember there are two answers to every question.

1 $x^2 - 2 = x$

2 $x^2 - 25 = 0$

3 $3x^2 = 48$

4 $12x = x^2$

5 $x^2 = 6x - 9$

6 $x^2 + 10x + 9 = 0$

7 $x^2 - 3x - 18 = 0$

8 $x^2 - 9x = 22$

9 $15x = 5x^2$

10 $32 = x^2 + 4x$

11 $x^2 - 18x + 45 = 0$

12 $x^2 - 13x + 42 = 0$

13 $x^2 = 11x + 42$

14 $x^2 = 11x - 28$

15 $x^2 = 3x + 28$

16 $x^2 + 25 = 10x$

17 $x^2 + 7x - 60 = 0$

18 $x^2 + 7x - 8 = 0$

19 $x^2 - 25 = 24x$

20 $4x^2 - 100 = 0$

15D

Solve the following equations. Remember there are two answers to every question.

1 $x^2 - 7x + 10 = 0$

2 $x^2 - 13x + 12 = 0$

3 $x^2 - x = 12$

4 $x^2 = 12x$

5 $3x^2 = 12x$

6 $3x^2 = 12$

7 $x^2 = 50 - 5x$

8 $5x^2 = 50 - 15x$

9 $x^2 + 15x + 26 = 0$

10 $x^2 - 7x = 30$

11 $x^2 - 7x = 18$

12 $x^2 - 13x + 30 = 0$

13 $x^2 - 13x - 30 = 0$

14 $45 = 5x^2$

15 $2x^2 - 5x = 3$

16 $8x^2 = 1 + 2x$

17 $7x + 15 = 2x^2$

18 $16x^2 - 9 = 0$

19 $9x^2 - 12x + 4 = 0$

20 $8x^2 + 5x = 3$

15E

1 The length of a rectangle is 5 cm more than its width. If the area is 84 cm², by writing a suitable equation find the dimensions of the rectangle.

2 The width of a rectangle is 3 cm less than half its length. The area of the rectangle is 108 cm². Calling the length $2x$, write an equation and hence find the dimensions of the rectangle.

3 The product of two numbers is 168 and their sum is 26. Write a suitable equation, and find the two numbers.

4 The area of a triangle is 27 cm². If the height is 3 cm more than the base, find the length of the base.

137

5 The sum of a number and its square is 72. Make up a suitable equation and find the number.

6 In a right-angled triangle the hypotenuse is 15 cm. If the other two sides differ by 3 cm find the length of the shorter side.

7 The shape shown is formed by a square with an isosceles triangle of height 6 cm drawn on one side. If the total area is 70 cm², write an equation and find the length of the side of the square.

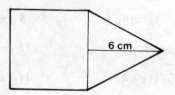

* One solution of your equation is negative, but it has a meaning. Can you see what the meaning is?

8 The width of a cuboid of height 5 cm is 2 cm less than its length, and its volume is 240 cm³. Find the length of the cuboid.

9 Three times a number subtracted from twice its square is the same as three times the number added to its square. What is the number?

10 The sum of the squares of two consecutive positive numbers is 265. Find the numbers.

11 The diameter of a circular dinner plate is 16 cm greater than the diameter of the top of a circular cup. The area of the plate is nine times the area of the top of the cup. If the radius of the cup is r cm, what is the radius of the plate? What are the two areas? Write an equation in r, solve it, and find the diameter of the plate.

12 Write down an expression for the total surface area of a cylinder of height h with two closed ends of radius r. If the total surface area is 96 π cm² and the height is 8 cm, what is the radius of the base?

(*Hint* In your final equation π and another number both cancel out, leaving a simple equation.)

13 What is the total surface area of a cone of slant height L and base radius r? If the total surface area is 36 cm² and the slant height is 4 cm, taking π as 3 find the radius of the base.

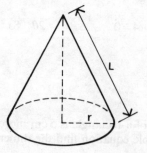

14 Repeat question 13 for a cone of slant height 7 cm and a total area of 90 cm².

15 One side of a rectangle is 11 cm longer than the other and a semi-circle is drawn on the shorter side. If the total area is 176 cm², find the dimensions of the rectangle. Take π as 3.

(*Hint* Take the short side as $2x$. The final equation will cancel down and give easy factors.)

16 A race track is in the shape of a rectangle with a semi-circle on each of the shorter sides. The length of the rectangle is 35 m greater than its width. The total area is 1·4 hectares (14 000 square metres). Taking π as 3, find the dimensions of the track.

(*Hint* Take the shorter side as $2r$.)

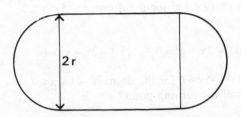

17 The area of a rectangle is three times the area of a square. One side of the rectangle is 2 cm longer than the side of the square and the other side is 4 cm longer. Find the length of the side of the square.

18 Repeat question 17 with the sides of the rectangle respectively 3 cm and 4 cm less than the sides of the square, and the area of the square double the area of the rectangle.

19 A firm makes rubber balls of various sizes. The retail price of these is $10x^2$ pence, where x is the radius of the ball in cm. The difference between the radii of two balls is 4 cm, and the larger ball costs nine times as much as the smaller. What does each cost?

20 If in question 19 the difference of the radii is 3 cm and the two balls together cost £4.50, what is the radius of each?

15F   The Graphical Solution of Quadratic Equations

1 Draw the graph of $y=x^2-2x$ for the domain $-3\leqslant x\leqslant 5$. (To do this, first complete the table below. This will give you the value of y for each integer value of x in the domain. Plot the points and draw the graph.)

x	-3	-2	-1	0	1	2	3	4	5
x^2 $-2x$	9 6								
y	15								

Use your graph to solve $x^2-2x=0$.
Draw the line $y=8$ and hence solve the equation $x^2-2x=8$.

2 Draw the graph of $y=x^2+2x-3$ for the domain $-5\leqslant x\leqslant 3$.
Use your graph to solve $x^2+2x-3=0$.
Write down the two values of x for which $y=-3$ on this curve.
These two x values are the solutions of a quadratic equation. Write down this equation.

3 Draw the graph of $y = x^2 - 4x + 4$ for the domain $-2 \leqslant x \leqslant 6$.
Solve the equation $x^2 - 4x + 4 = 0$.
Draw the line $y = 9$ and hence solve the equation $x^2 - 4x - 5 = 0$.
By drawing a second line solve $x^2 - 4x = 0$.
What is the equation of this second line?

4 Draw the graph of $y = 15 - 2x - x^2$ for the domain $-5 \leqslant x \leqslant 3$.

Use your graph to solve:

a) $15 - 2x - x^2 = 0$ b) $8 - 2x - x^2 = 0$ c) $3 - 2x - x^2 = 0$

5 Draw the graph of $y = x^2 - 5x + 6$ for the domain $-1 \leqslant x \leqslant 7$.
What are the co-ordinates of the turning point?

Use your graph to solve:

a) $x^2 - 5x + 6 = 0$ b) $x^2 - 5x = 0$ c) $x^2 - 5x - 4 = 0$

Check your answer to c and find out how accurate your graph is.

6 Draw the graph of $y = 2x^2 - 3x - 5$ for the domain $-3 \leqslant x \leqslant 4$.
What is the minimum value of y?

Use your graph to solve:

a) $2x^2 - 3x - 5 = 0$ b) $2x^2 - 3x = 0$ c) $2x^2 - 3x - 15 = 0$

7 Draw the graph of $y = 3 - 4x - 4x^2$ for the domain $-4 \leqslant x \leqslant 3$.
What is the maximum value of y?

Use your graph to solve:

a) $3 - 4x - 4x^2 = 0$ b) $4x^2 + 4x = 0$ c) $15 - 4x - 4x^2 = 0$
d) $31 - 4x - 4x^2 = 0$

8 Draw the graph of $y = 4 + 2x - x^2$ from $x = -3$ to $+5$.

a) What is the maximum value of y?

Use your graph to solve the following equations:

b) $4 + 2x - x^2 = 0$ c) $7 + 2x - x^2 = 0$ d) $11 + 2x - x^2 = 0$

Give your answers to 1 d.p.

9 Draw the graph of $y = x^2 + 3x - 6$ from $x = -5$ to $+2$.

a) Give the co-ordinates of the turning point.

Use your graph to solve the equations:

b) $x^2 + 3x - 6 = 0$ c) $x^2 + 3x - 2 = 0$ d) $x^2 + 3x + 1 = 0$

Give your answers to 1 d.p.

10 Draw the graph of $y = x^2 - 8x + 10$ from $x = -3$ to $+5$.

a) What is the minimum value of y?

Use your graph to solve the following equations:

b) $x^2 - 8x + 10 = 0$ c) $x^2 - 8x + 6 = 0$ d) $x^2 - 8x - 2 = 0$

Give your answers to 1 d.p.

11 A stone is hurled vertically upwards into the air. After t seconds its distance s metres above its starting point is given by the equation $s = 40t - 5t^2$. When the stone has returned to its starting point $s = 0$. After what time does this occur?
Find how far it is above its starting point after 1, 2, 3,...7 seconds, i.e. find s for $t = 1$, 2, 3,...7 seconds.
Using 2 cm to 1 sec on the horizontal axis, and 2 cm to 10 m on the vertical axis, draw a graph to show $s = 40t - 5t^2$.
From your graph find the answers to these questions:

a) How long does the stone take to reach its maximum height?
b) For how long is it above the height of 60 m?
c) For how long is it below the height of 35 m?

12 If a boy catapults a stone vertically upwards from the edge of a cliff, the distance s metres of the stone above the ground after t seconds is given by the equation $s = 25t - 5t^2$.
Find values of s for the first 7 seconds of its motion and draw a graph to show this relationship between s and t.
From your graph answer these questions:

a) What is the maximum height reached and how long does it take to reach this height?
b) After what times is the stone 20 m away from the point of projection?
c) If after 6·25 seconds the stone strikes the rocks at the foot of the cliff, how far above the rocks was the boy standing?

✳ *13* A farmer has 110 m of fencing with which to make a rectangular pen for his sheep. Let x metres be the length of one side and A square metres the area enclosed.
Write an equation to relate A and x.
Work out values of A for x between 5 and 50.
Draw a graph to show the relationship between x and A and from it answer the following questions:

a) What is the maximum area?
b) What are the dimensions of the rectangle which has maximum area?
c) What are the dimensions of the pen when the area enclosed is 700 m²?

16 The Geometry of the Circle

16A Chords

Note i) In each of the following questions *O* marks the centre of the circle.

ii) Where answers are not exact give lengths to 1 d.p.

1 *AB* is a chord of the circle.

a) What kind of triangle is *OAB*?

b) If *OP* is perpendicular to the chord, what do you know about *AP* and *PB*?

c) What do you know about the angles *AOP* and *POB*?

2 *LM* is a chord of the circle.
If *Q* is the mid-point of *LM* what do you know about *OQ* and *LM*?
What do you know about the two triangles *QLO* and *QMO*?

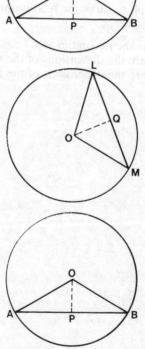

3 *AB* is a chord of length 8 cm in a circle of radius 5 cm. Find the perpendicular· distance of the chord from the centre, i.e. *OP* in the diagram.

4 Find the length of the chord which is 4 cm from the centre of a circle of radius 5 cm.

5 *XY* is a chord and *P* is the mid-point of *XY*.

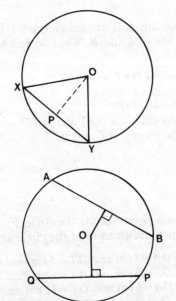

  a) If $OX = 13$ cm and $OP = 12$ cm find *XY*.
  b) If $OX = 10$ cm and $OP = 7$ cm find *XY*.
  c) If $OX = 12$ cm and $XY = 16$ cm find *OP*.
  d) If $OX = 25$ cm and $XY = 48$ cm find *OP*.
  e) If $XY = 18$ cm and $OP = 7$ cm find *OX*.
  f) If $XY = 26$ cm and $OP = 6$ cm find *OX*.

6 *AB* is a chord of length 10·8 cm whose perpendicular distance from the centre of the circle is 3 cm. Find the radius of the circle. If *PQ* is a second chord in the circle and is 4·5 cm from the centre, find the length *PQ*.

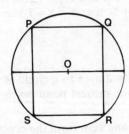

7 *AB* and *CD* are two parallel chords in a circle of radius 8 cm. If $AB = 10$ cm and $CD = 13·6$ cm, find the perpendicular distance between the two chords. (Two answers.)

8 A chord of length 9 cm is drawn inside a circle of radius 9 cm. How far is the chord from the centre? How many such chords can you draw in the circle if there is no overlapping?

9 In a circle of radius 25 cm, two parallel chords are drawn, both on the same side of the centre. If one is of length 40 cm and the other of length 30 cm, how far apart are they?
If one had been drawn on each side of the centre, how far apart would they then have been?

10 *PQRS* is a rectangle drawn in a circle of radius 10 cm. If the mid-points of *PS* and *QR* are each 2 cm from the ends of the diameter on which they lie, find the dimensions of the rectangle.

16B   Tangents

Note i) In each of the following questions *O* marks the centre of the circle.
  ii) Where answers are not exact give lengths to 1 d.p.

1 *AB* is a chord which is perpendicular to the radius *OR*. It is allowed to move parallel to itself between *O* and *R*. In one extreme position it becomes the diameter (*CD*). In the other extreme position it becomes a tangent to the circle (*EF*). What can you deduce about the angle between the tangent at *R* and the radius *OR*?

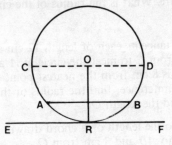

2 Two tangents are drawn from T to meet the circle at A and B. What do you know about

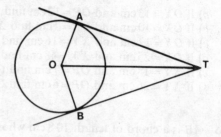

a) angles OAT and OBT?
b) lengths OA and OB?
c) the line OT?
d) the triangles OAT and OBT?
e) the lengths AT and BT?

3 T is a point outside the circle. From T a tangent is drawn to meet the circle at A.

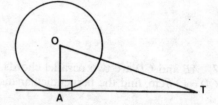

a) If $OA=3$ cm and $AT=5$ cm find OT.
b) If $OA=6$ cm and $AT=8$ cm find OT.
c) If $OA=6$ cm and $OT=12$ cm find AT.
d) If $OA=10$ cm and $OT=15$ cm find AT.
e) If $AT=7$ cm and $OT=12$ cm find OA.
f) If $AT=4.5$ cm and $OT=8$ cm find OA.

4 XY is a tangent drawn from the point Y outside the circle meeting the circle at X.

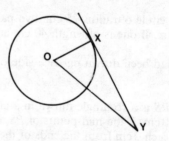

a) If $OX=1.2$ cm and $XY=3$ cm find OY.
b) If $OX=2.5$ cm and $OY=6$ cm find XY.
c) If $XY=8$ cm and $OY=11$ cm find OX.
d) If $OX=5$ cm and $OY=12$ cm find XY.
e) If $OX=4.2$ cm and $XY=7.5$ cm find OY.

5 A tangent is drawn to a circle of radius 7 cm from a point outside the circle which is 6 cm from the nearest point on the circumference. Find the length of the tangent.

6 From a point 9 cm from the centre of a circle, a tangent is drawn to the circle. If the radius is 5·8 cm, find the length of the tangent.

7 A tangent of length 5·5 cm is drawn to a circle from a point which is 8·5 cm from the centre. What is the radius·of the circle?

8 Two tangents each of length 12 cm are drawn from C to meet the circle at A and B. If C is 8 cm from the nearest point on the circumference, find the radius of the circle and the length OC.

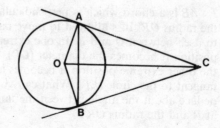

Find also the length of a chord drawn parallel to AB and 3 cm from O.

144

9 *EA* and *EC* are tangents to the circle at *A* and *D* respectively. If the radius of the circle is 5 cm, *C* is 13 cm from the centre and the length *CE* is 19·5 cm, find the length *EA* and the distance of *E* from *O*.

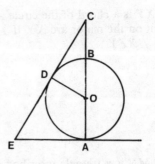

✱10 A circle of radius 8 cm is inscribed in an equilateral triangle. If the centre of the circle is ⅓ of the perpendicular height from the base of the triangle, find the length of the sides of the triangle. If two of the tangents touch the circle at the points *A* and *B*, find the length of the chord *AB*.

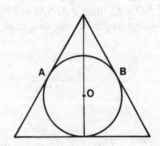

16C Angle Properties

Note In each of the following questions *O* marks the centre of the circle.

1 In the figure *AC* and *CB* are chords of the circle. The line joining *C* to *O* is produced to *D*.
If ∠*ACO* = 20° and ∠*OCB* = 35°, calculate ∠*AOD* and ∠*DOB*.
Write down ∠*ACB* and ∠*AOB* and the relationship between these two.

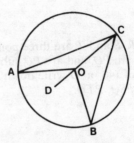

2 Repeat question 1 for

a) ∠*ACO* = 28°, ∠*OCB* = 37°   b) ∠*ACO* = 30°, ∠*OCB* = 41°
c) ∠*ACO* = 25°, ∠*OCB* = 18°.

3 Using the relationship which you found in the previous questions between the angle subtended by *AB* at the centre and the angle subtended by *AB* at the circumference, calculate the values of *a, b, c, d, e, f, g* in the following diagrams:

(*Note* The angle 'subtended' by the line *PQ* at the point *R* is ∠*PRQ*.)

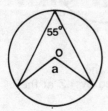
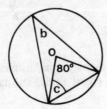
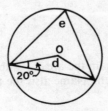
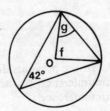

4 XY is a chord of the circle and Z is any point on the major arc XY. If $\angle OXY = 39°$ find $\angle XZY$.

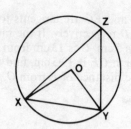

5 LMN is a triangle inscribed in a circle. If $\angle LNO = 32°$ and $\angle ONM = 25°$, calculate all the other angles in the diagram.

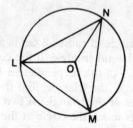

6 AB is a chord of the circle and C is a point on the major arc AB such that BC is parallel to AO.
If $\angle AOB = 66°$ calculate $\angle OAC$.
Find also the angles OBA and BAC.

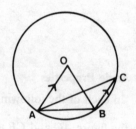

7 If R, S and T are three points on the major arc PQ, and arc PQ subtends an angle of $140°$ at O write down the angles PRQ, PSQ, PTQ.

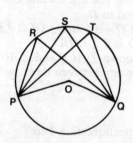

8 If X, Y and Z are three points on the major arc AB, and arc AB subtends an angle of $42°$ at X write down the angles AYB and AZB.

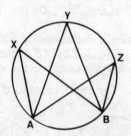

9 In the previous diagram the arc XY subtends two equal angles at the circumference. Name them. Name also the two equal angles subtended by YZ at the circumference.

10 Copy the previous diagram and draw in three lines to make three equal angles subtended by the chord *AX* at the circumference. Name the lines and the equal angles. The diagram should now contain two equal angles subtended by the chord *BZ* at the circumference. Name these angles.

11 Calculate the values of *a, b, c, d, e, f, g, h* in the following diagrams:

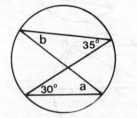 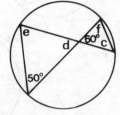 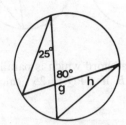

12 *PQ* and *RS* are two parallel chords of the circle. If ∠ *PQR* = 42°, calculate ∠ *PSR*, ∠ *QRS*, ∠ *QPS*. What do you know about the two triangles in the diagram?

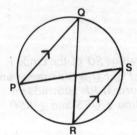

13 *ABC* is an isosceles triangle in which *AB* = *AC*. The triangle is inscribed in a circle. Copy the diagram and mark in the equal angles and the equal lengths. What do you think this tells you about angles subtended at the circumference by equal chords?

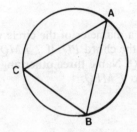

14 If the two arcs *LM* and *MN* are equal in length and ∠ *LKM* = 32°, find ∠ *MKN*. Calculate the three angles of triangle *LMN*.

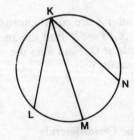

15 *AB* is a diameter of the circle. What angle does the arc *AB* subtend at the centre of the circle? If *P* and *Q* are two points on the circumference of the circle, what do you know about the angles *APB* and *AQB*?

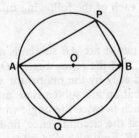

16 Calculate the values of $a, b, c, d, e, f, g, h, i$ in the following diagrams:

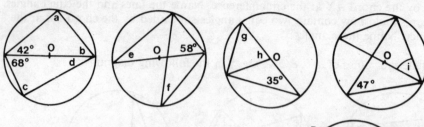

17 PQ is a chord which is parallel to the diameter AB. If $\angle PQA = 25°$, calculate $\angle QBA$.

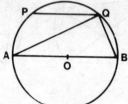

18 The radius SO of the circle is parallel to the chord PQ. RQ is a diameter and RS a second chord. If PR subtends an angle of $46°$ at Q, find $\angle QRS$ and $\angle RSO$.

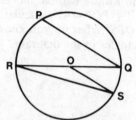

19 LM is a diameter of the circle which is parallel to the chord PQ. If $\angle LMQ$ is $63°$ find $\angle MPQ$. Name three other angles which are equal to $\angle MPQ$.

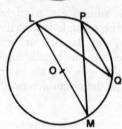

20 XY and YZ are equal chords of the circle. If $\angle XZY = 52°$ find the angle which YZ subtends at O. Find also the angle which XZ subtends at Y.

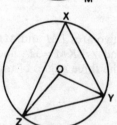

16D Cyclic Quadrilaterals

Note In each of the following questions O marks the centre of the circle.

1 If the minor arc AB subtends an angle of $160°$ at O, find the reflex angle AOB which is subtended at O by the major arc AB. Using the fact that an arc subtends an angle at the centre which is twice the angle which it subtends at the circumference, find $\angle APB$ and $\angle AQB$. Find also $\angle APB + \angle AQB$.

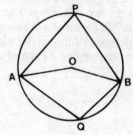

2 Repeat question 1 when   a) $\angle AOB = 140°$ b) $\angle AOB = 128°$
c) $\angle AOB = 154°$ d) $\angle AOB = 135°$.

3 The quadrilateral $APBQ$ in questions 1 and 2 is called a 'cyclic quadrilateral' because the four vertices all lie on a circle. What do you deduce about the opposite angles of a cyclic quadrilateral?

4 $ABCD$ is a cyclic quadrilateral in which $\angle A = 136°$ and $\angle B = 76°$.
Calculate $\angle C$ and $\angle D$.

5 Find the values of $a, b, c, d, e, f, g, h, i, j, k$ in the following diagrams:

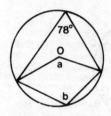 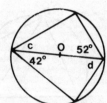 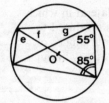 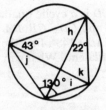

6 PQ is a diameter of the circle. R and S are two points on the circumference as shown in the diagram. If $\angle QOR = 56°$ find $\angle OQR$ and $\angle PSR$.

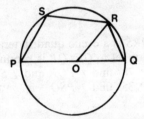

7 LMN is an isosceles triangle where $LM = LN$. If $\angle MLN = 64°$ calculate $\angle LKN$.

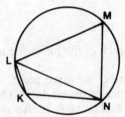

8 The side CD of the cyclic quadrilateral $ABCD$ is produced to E. If $\angle EDA = 130°$ find $\angle CBA$.

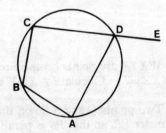

9 Using the diagram of question 8, find

a) $\angle CBA$ if $\angle EDA = 105°$,   b) $\angle EDA$ if $\angle CBA = 117°$,

c) $\angle EDA$ if $\angle CBA = 88°$.

What conclusion do you come to about an interior angle of a cyclic quadrilateral and the exterior opposite angle?

10 Find the values of $a, b, c, d, e, f, g, h, i, j$ in the following diagrams:

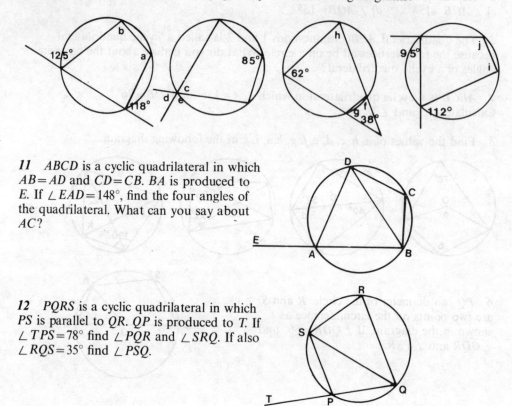

11 ABCD is a cyclic quadrilateral in which $AB = AD$ and $CD = CB$. BA is produced to E. If $\angle EAD = 148°$, find the four angles of the quadrilateral. What can you say about AC?

12 PQRS is a cyclic quadrilateral in which PS is parallel to QR. QP is produced to T. If $\angle TPS = 78°$ find $\angle PQR$ and $\angle SRQ$. If also $\angle RQS = 35°$ find $\angle PSQ$.

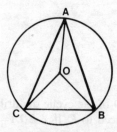

16E

Note In each of the following questions O marks the centre of the circle.

1 A, B and C are three points on the circumference of the circle. If AB subtends 130° at O, find $\angle ACB$. If also BC subtends 90° at the centre find the angles of triangle AOC.

2 WXYZ are points on the circumference of a circle such that $XZ = XY$ and $\angle YXZ = 72°$. Calculate $\angle XZY$ and $\angle XWY$.

3 Two points Q and S lie on the circumference of a circle on opposite sides of the diameter PR, so that PS is parallel to QR. If $\angle RPS = 52°$ find angles PRQ and QPR. What do you know about QS?

4 The chord LM in a circle is parallel to the diameter PR. If $\angle PLM = 107°$ calculate angles PRM, RPM and PML.

5 ABCD is a cyclic quadrilateral such that $BC = BD$. DA is produced to E. If $\angle EAB = 68°$ and $\angle ABD = 32°$ calculate the angles of the quadrilateral ABCD.

6 *ABCD* is a cyclic quadrilateral with *AB* and *DC* produced to meet at *P* and *DA* and *CB* produced to meet at *Q*. If $\angle CBP = 50°$ and $\angle APD = 65°$, calculate the angles of triangle *CDQ*. Find also the angles of triangle *ABQ*. What do you know about these two triangles? Name another pair of triangles which are related in the same way.

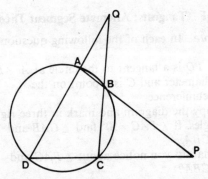

7 Two chords *PQ* and *RS* are produced to meet outside the circle at *T*. If $\angle PQR = 58°$ and $\angle PTR = 24°$, calculate angles *PSR, PST, TPS* and *TRQ*. Name two pairs of similar triangles in this diagram.

8 Chord *PQ* of a circle is parallel to the radius *OA*. If $\angle PQO = 58°$ find angles *QOA* and *QPA*.

9 Chord *XY* of a circle subtends an angle $a°$ at *Z*, a point on the circumference. Express angles *XOY* and *OXY* in terms of a.

10 *ABCD* is a cyclic quadrilateral in which $AC = CD$. If $\angle ACD = 2x°$ express angles *CDA* and *ABC* in terms of *x*.

11 Chord *ST* of a circle and the diameter *PQ* are both produced to meet at *R*. If $SQ = QR$ and $\angle SQP = 2y$, write expressions in terms of *y* for angles *SOP* and *SRP*.

12 *XYZ* is an isosceles triangle in which $ZX = ZY$. It is inscribed in a circle, *W* and *U* being two other points on the circumference as shown. If $\angle ZYX = p°$ write expressions for angles *XWY* and *XUY* in terms of *p*.

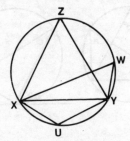

13 *PQR* is an isosceles triangle in which $PR = PQ$. It is inscribed in a circle with *QS* as diameter. If $\angle PRQ = x°$ and $\angle SQR = y°$ find an equation to show the relationship between *x* and *y*. (Two answers.)

14 The diameter *AB* of a circle is produced to meet the chord *CD* produced at *E*. If $\angle CAB = 34°$ and $\angle BCE = 28°$ prove that $CB = BE$.

15 *AB* is a diameter and *DE* a chord of the circle. When both are produced they meet at *C*. If $\angle ODE = 54°$ and $\angle OEA = 18°$, prove that $AE = EC$.

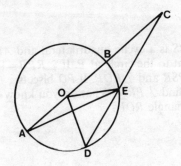

151

16F   Tangents: Alternate Segment Theorem

Note   In each of the following questions *O* marks the centre of the circle.

1   *PQ* is a tangent to the circle at *A*. *AB* is
a diameter and *C* is a point on the
circumference.
Copy the diagram and mark in three right
angles. If $\angle PAC = 50°$ find $\angle CAB$ and
$\angle CBA$.
What do you notice about $\angle PAC$ and
$\angle CBA$?
Draw on your diagram two other angles
each equal to *PAC*.

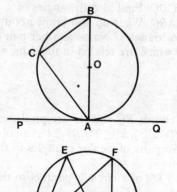

2   *AC* is a tangent to the circle at *B*. *BE* is
a diameter and *D* and *F* are points on the
circumference.

a) If $\angle CBD = 43°$ find $\angle BED$ and $\angle BFD$.
b) If $\angle CBD = 56°$ find $\angle BED$ and $\angle BFD$.
c) If $\angle EBD = 40°$ find $\angle CBD$ and $\angle BFD$.
d) If $\angle DEB = 39°$ find $\angle CBD$ and $\angle BFD$.

What can you say about angles *CBD* and
BFD?

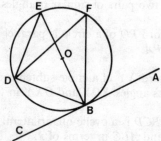

3   Calculate the angles *a, b, c, d, e, f, g* in the following diagrams:

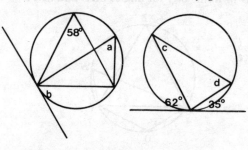

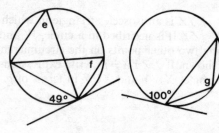

4   *TS* is a tangent to the circle at *A*.
B and *C* are two other points on the
circumference. If $\angle TAC = 40°$ find
$\angle ABC$ and $\angle AOC$.

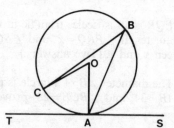

5   *PQRS* is a cyclic quadrilateral and *AB* is
a tangent to the circle at *P*. If $\angle RPB = 110°$
find $\angle PSR$ and $\angle RQP$. If *PQ* bisects
$\angle BPR$ find $\angle PRQ$. What do you know
about triangle *RQP*?

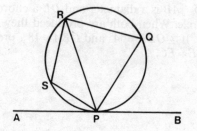

6 Two tangents XY and XZ meet the circle at A and B respectively. C is a point on the major arc AB. If $\angle ABX = 63°$ find $\angle BCA$ and $\angle BAX$.
Name two other pairs of equal angles.

7 AB, a diameter of the circle, is produced to C and EC is a tangent to the circle at D. If $\angle ABD = 53°$ find angles BAD, BDC and DCB.

8 Two tangents XY and XZ meet the circle at A and B respectively. C is a point on the circumference such that CB is parallel to YX. If $\angle AXB = 68°$ find angles ZBC and BAC.

9 The diameter PQ of a circle is produced to T, and TR is a tangent which meets the circle at S. If $\angle STP = 36°$ find angles SOQ, SPQ and PSR.

10 The circle inscribed in triangle ABC touches the sides at X, Y and Z. If $\angle YAZ = 62°$ and $\angle YZX = 48°$, find angles ACB and ABC.

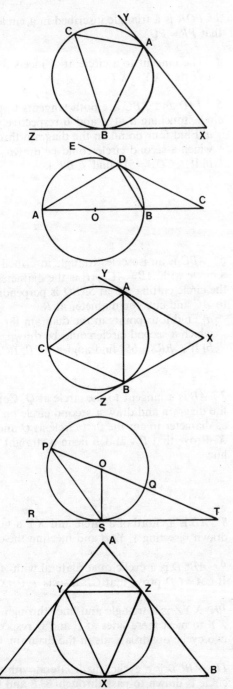

16G Miscellaneous

Note In each of the following questions O marks the centre of the circle.

1 Two tangents AB and AC are drawn to meet a circle at P and Q respectively. R is a point on the major arc PQ such that the chord QR is parallel to AB. Prove that $PR = PQ$.

2 The tangent to the circle at S cuts the chord PQ (produced) at R. Prove that $\angle SQR = \angle PSR$.

3 *PQR* is a triangle inscribed in a circle. If the tangent at *P* is parallel to *RQ*, prove that *PR = PQ*.

4 The tangent to a circle at *X* meets the diameter *YZ* (produced) at *T*. If *XZ = ZT* prove that *XY = XT*.

5 *TAD* and *TBC* are both tangents to the circle, touching it at *A* and *B* respectively.
a) Find four points in the diagram through which a second circle could be drawn.
b) If ∠*COB* = 51° find ∠*BTA*.

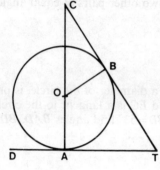

6 *ABC* is an isosceles triangle inscribed in a circle with *AB = AC*. *AP* is the diameter of the circle cutting *BC* at *S*. *BQ* is perpendicular to *AC* and cuts the diameter at *R*.
a) Find four points in the diagram through which a second circle could be drawn.
b) If ∠*BRS* = 65° find angles *ACB*, *BCP* and *BAP*.

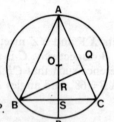

7 *AB* is a tangent to the circle at *Q*. Copy the diagram and draw a second circle on *QB* as diameter to cut the first circle at *Q* and *X*. Prove that *PX* and *B* lie in a straight line.

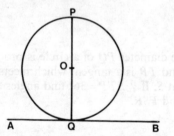

8 *AB* is a chord of a circle and *XY* a tangent to the circle at *A*. If a second chord is drawn bisecting ∠*YAB* and meeting the circle again at *C*, prove that *AC = CB*.

9 *ABCD* is a cyclic quadrilateral with *AB* and *DC* produced to meet at *X*. If *CA = CD* prove that *CB* bisects ∠*DBX*.

10 *XYZ* is a triangle and lines through *Y* and *Z* are drawn perpendicular to *XZ* and *XY* to meet these sides at *P* and *Q* respectively. If these two altitudes meet at *R*, name two cyclic quadrilaterals in the diagram. Calculate ∠*RXP* given that ∠*XZY* = 56°.

11 *ABCD* is a cyclic quadrilateral with *AB* and *DC* produced to meet at *E*. A second circle is drawn to pass through *B*, *E* and *C*. If ∠*DAB* = 55° and ∠*ADC* = 70°, find the angles of triangle *BCE*. Name two pairs of equal lines in the diagram.

12 The larger circle in the diagram passes through the centre of the smaller circle and *ACD* and *AOB* are straight lines. Prove that *BD* is a diameter of the larger circle. If angle *COB* = *x*° find angles *CDB* and *CAB* in terms of *x*.

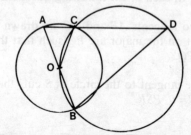

154

16H Questions Involving Trigonometry

Note i) In each of the following questions O marks the centre of the circle.

ii) Where answers are not exact, give lengths to 3 s.f. and angles to the nearest minute (or to $0.1°$ if using three-figure tables).

1 Two tangents are drawn from a point T outside a circle of radius 5 cm. The tangents are each of length 8·5 cm and meet the circle at A and B.

a) Calculate the angle between the tangents at T.
b) Calculate the angle between the two radii OA and OB.
c) Name four points through which a second circle can be drawn, giving reasons.
d) Find the radius of this second circle.

2 XYZ is an isosceles triangle in which $XY = XZ = 9$ cm and angle $X = 40°$. Lines are drawn through Y and Z at right angles to XZ and XY to meet these sides at P and Q respectively.

a) Name four points through which a circle can be drawn. Give reasons for your answer.
b) Find the length of the diameter of this circle.

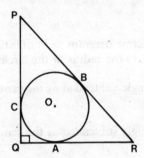

3 A circle of radius 2 cm is inscribed in a right-angled triangle PQR. If the circle touches the sides of the triangle at A, B and C and $QC = \frac{1}{2}CP$ and $QA = \frac{1}{3}AR$, find the length PR. Find also the angles AOB, BOC, COA.

4 Two tangents are drawn from X to meet a circle at A and B. If $XA = XB = AB = 6$ cm, find the radius of the circle and the distance of the chord AB from the centre O of the circle.

5 Two circles are drawn one inside the other, both having the same centre O. The radius of the inner one is 6 cm. A chord of the outer circle, of length 9 cm, is also a tangent to the smaller circle. Find the radius of the outer circle, and the angle the chord subtends at O.

6 Two circles, centres O_1 and O_2, are drawn touching a common tangent at A and B. The line joining their centres meets the tangent at C.

a) What do you know about triangles O_1AC and O_2BC?
b) If the radius $O_1A = 12$ cm, $AC = 16$ cm and $BC = 4$ cm, find the radius O_2B and the lengths O_1O_2 and O_1C.
c) Calculate the angle at C.

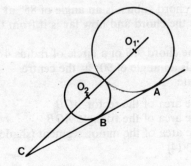

7 A circle is inscribed in an equilateral triangle. The triangle is then circumscribed by a second circle. If this second circle has a radius of 7 cm, find the radius of the inner circle and the length of the sides of the triangle, remembering that these sides are tangents to one circle and chords of the other circle.

8 In a vertical circle of radius 35 cm, if OA is vertical and OB is at 48° to OA, how high is B above the level of A? If $\angle AOC = 135°$ how high is C above the level of A?

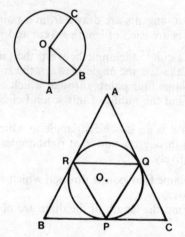

9 ABC is an isosceles triangle in which $AB = AC$ and angle $A = 44°$. A circle inscribed in the triangle touches the sides at P, Q and R. If $AQ = 5·5$ cm, find
a) RQ, b) $\angle AQR$, c) $\angle QPR$, d) PQ.

10 Using the same diagram as in question 9, if $AC = AB = 15$ cm and $\angle ACB = 65°$, calculate a) PC, b) the radius of the circle.

11 Find the angle subtended at the centre of a circle of radius 3 cm by a chord of length 5 cm.

12 Find the angle subtended at the centre of a circle of radius 9 cm by a chord of length 12 cm.

13 A chord of length 8 cm subtends an angle of 108° at the centre of a circle. Find the radius of the circle.

14 Two circles, centres O_1 and O_2, have a common chord PQ 10 cm long. If PQ subtends an angle of 90° at O_1 and 118° at O_2, find the distance O_1O_2.

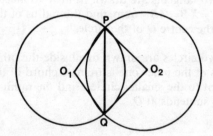

15 A chord subtends an angle of 85° at the centre of a circle of radius 9 cm. How long is the chord and how far is it from the centre of the circle?

16 The chord AB of a circle of radius 4 cm subtends an angle of 90° at the centre. Calculate

    a) the area of the sector AOB,
    b) the area of the triangle AOB,
    c) the area of the minor segment (shaded).
    $(\pi = 3·14)$

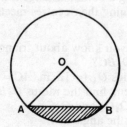

17 If the angle subtended by the chord AB at the centre of a circle of radius 10 cm is 60°, find

  a) the area of the sector AOB,
  b) the length of the chord AB,
  c) the distance of the chord from the centre,
  d) the area of the triangle AOB,
  e) the area of the minor segment.
  ($\pi = 3.14$)

18 The length of the arc of a segment is 5·5 cm and the radius of the circle is 7 cm. Find the angle which the arc subtends at the centre of the circle and the length of the chord which cuts off this arc. ($\pi = \frac{22}{7}$)

∗ 19 Find the area of the segment of a circle of radius 12 cm, if the chord which cuts off this segment subtends an angle of 80° at the centre. ($\pi = 3.14$)

∗ 20 Find the area of a segment of a circle of radius 10 cm which is a part of a sector of area $25\pi \text{ cm}^2$.

17 Linear Programming

17A Introduction

In questions 1–4, mark the x and y axes with values from 0 to 12.

1 Draw the lines $y=x$, $x=1$, $x+y=10$ and show by shading the region for which $y>x$, $x>1$, $x+y<10$. If x and y are whole numbers, dot in the solution set.

a) What is the largest possible value of x? What is the smallest possible value for y?
b) If the definition of the region is changed so that $x+y\leqslant 10$ and $y\geqslant x$, what is the largest possible value for x now?

2 Draw the lines $x+2y=10$, $y=2x$, $x=8$, $y=7$ and show, by shading, the region for which $x+2y\geqslant 10$, $y\leqslant 2x$, $x\leqslant 8$, $y\leqslant 7$. If x and y are whole numbers, what is the largest possible value of y? What is the smallest? Find the largest possible value of $x+y$.

3 Show graphically the possible values of x and y where $y\geqslant\frac{1}{2}x$, $x+2y>8$, $y>0$, $x+y\leqslant 8$. If x and y are whole numbers, dot in the solution set and find the smallest possible value of $x+y$.

4 Show by shading the region for which $2x+3y\geqslant 12$, $y<2x+1$, $x<5$, $3x+4y\leqslant 36$. List all the points in the region whose x and y co-ordinates are whole numbers. Do any of them lie on the line $3x+4y=36$?

In questions 5 and 6, mark values from 0 to 60 on both axes.

5 Show by shading the region for which $x+y\leqslant 50$, $y\leqslant 2x$, $3x+10y\geqslant 150$. If x and y are multiples of 10, find the possible totals for $x+y$.

6 Show graphically the possible values of x and y if: $y\leqslant 50$, $x\leqslant 45$, $y\leqslant 3x$, $4x+5y\geqslant 200$. What range of values of y is allowed?

In questions 7 and 8, mark values from 0 to 120 on both axes.

7 Show by shading the region $y\geqslant\frac{1}{2}x$, $x<60$, $2x+y\geqslant 100$, $2x+3y\leqslant 240$. If x and y are multiples of 10, give the smallest and largest possible values of $x+y$.

8 Show graphically $y\geqslant 0$, $y<x$, $3x+y\geqslant 120$, $x\leqslant 90$, $4x+7y\geqslant 280$. What is the range of possible values of x?

17B Linear Programming

1 x and y are two positive integers (i.e. not including zero).

a) If their sum does not exceed 7, express this as an inequality.
b) If the number obtained by taking y from x is not less than 1, express this as another inequality.
c) Draw a graph of the lines $x + y = 7$ and $x - y = 1$. Shade lightly the region in which x and y must lie. Mark also (with a heavy dot) the grid points in this region, i.e. the points at which both x and y are integers.
d) Find the grid point at which the sum of $x + 2y$ is a maximum.
e) Find also the grid point at which the sum of $2x + y$ is a maximum.
f) Do these grid points i) both lie on a boundary, ii) both lie at a node?

2 The owner of a shop employs f full-time assistants and p part-time assistants.

a) At no time does he employ more than seven assistants altogether. Express this as an inequality in f and p.
b) He arranges that the number of full-time assistants is at least 1 greater than the number of part-time assistants. Express this as an inequality in f and p.
c) Draw a graph to illustrate these two inequalities. Shade lightly the area in which f and p must lie.
d) Mark the grid points in the shaded area. Why?
e) A full-time assistant is paid twice as much as a part-time assistant. Find from your graph the combination of p and f for which the owner pays out the largest sum in wages. (*Hint*   Find the node at which $2f + p$ is a maximum.)
f) Can you see any similarity between questions 1 and 2?

3 A baker makes two types of cakes, Moorish and Swish.

a) Each Moorish requires 400 cm^2 of oven space, and each Swish 600 cm^2. If he makes m of the first and s of the second each day, how much oven space is required altogether?
b) The total oven space available is $10 \cdot 8 \text{ m}^2$. Express this as cm^2.
c) Write an inequality concerning oven space. Cancel this down to make it as simple as possible.
d) He has one assistant working full time at cake making, and the average time this assistant spends on one Moorish and one Swish is 3 mins and 2 mins respectively. What times does he spend on m Moorish and s Swish?
e) If this assistant works $8\frac{1}{2}$ hours a day (including overtime), how many minutes a day is this?
f) Using your answers to d and e write an inequality about 'cake-making time'.
g) Draw a graph illustrating your two inequalities and shade the region in which m and s must lie.
h) If the profit on one Moorish is 12p and on one Swish 10p, find from your graph the combination of m and s which gives the greatest profit.
(*Hint*   Find the value of $12m + 10s$ at each node in turn.)

4 The owner of a café is buying new tables.

a) If he buys b bigger tables each taking 3 m^2 of floor space (including room for chairs) and s smaller tables each taking 2 m^2 of floor space, what floor space is required altogether?
b) If the available floor space is 110 m^2 write an inequality in b and s.
c) What is the total cost of b bigger tables at £20 each and s smaller tables (export rejects) at £5 each?
d) If the total cost of the tables must not exceed £800 write another inequality in b and s. (Cancel this down as much as possible.)

e) Represent your two inequalities on a graph and shade lightly the region in which *b* and *s* must lie.

f) If the bigger tables seat 4 and the smaller 2 find from your graph the combination of *b* and *s* which gives the largest number of seats.

5 A contractor accepts an urgent contract to move earth. As all his vehicles are engaged on other work he buys new ones. The most suitable are Topnotch Trucks of which he buys *t* and Wonder Waggons of which he buys *w*.

a) If he is unable to get more than 20 Topnotch Trucks, write an inequality in *t*.

b) If he is unable to get more than 30 Wonder Waggons, write an inequality in *w*.

c) If each Topnotch Truck can take 20 tonnes of earth and make 8 journeys a day, how much earth can be moved in a day by *i*) one truck, *ii*) *t* trucks.

d) If each Wonder Waggon can take 12 tonnes of earth and make 10 journeys a day, how much earth can be moved in a day by *i*) one waggon, *ii*) *w* waggons.

e) How much earth can be moved altogether in one day?

f) If the contract is to move at least 4400 tonnes of earth a day, write down an inequality in *t* and *w* expressing this fact.

g) Draw a graph and shade the area in which *t* and *w* must lie.

h) If a Topnotch Truck costs £30 000 and a Wonder Waggon £25 000 what combination of *t* and *w* will make the cost a minimum? (*Hint* Find the value of $30t + 25w$ at each node of the shaded area.)

6 If in question 5 the contractor decides to use as many trucks and waggons as possible ignoring the cost, but he can only get 45 drivers (and therefore can only use 45 trucks and waggons) write another inequality in *t* and *w*. Copy the graph from question 5 and add another line representing this fresh inequality. Shade the region in which *t* and *w* must now lie. Which combination of *t* and *w* will enable him to get the job done most quickly, i.e. to get the most earth moved each day? (*Hint* Find the number of tonnes of earth moved in a day at each node, using the results of 5*e*.)

7 An office manager in a rapidly expanding firm is buying desks for a new office. He chooses two types, *A* and *B*.

a) If type *A* costs £60 each and *B* £100, and he buys *a* of *A* and *b* of *B*, what is the total cost?

b) If he is not allowed to spend more than £3000 altogether on desks, write down an inequality concerning costs.

c) If each desk of type *A* requires 5 m² of space, including the surround, and *B* 4 m², how much space will be required by the desks?

d) The total floor area available is 200 m². Write down an inequality concerning space.

e) On graph paper, draw a graph illustrating the two inequalities and shade lightly the area in which *a* and *b* must lie.

f) From your graph find the values of *a* and *b* which will make the total number of desks a maximum. (*Hint* Find the value of $a + b$ at or near each node of the shaded area. *a* and *b* must be whole numbers.)

8 The owner of a newly formed car hire organisation is buying two types of cars, *P* which is small and cheap to run but wears out quickly, and *Q* which is larger and costs more to run but lasts longer.

a) If he buys *p* of *P* at £2500 each and *q* of *Q* at £3500 each, what is the total cost?

b) If he decides not to spend more than £70 000 on cars, write down an inequality concerning cost.

c) If a car of type *P* requires 9 m² of garage space and a car of type *Q* 18 m², what is the total garage space required?

160

d) If no more than 315 m^2 of garage space is available, write down an inequality concerning garage space.

e) Using these two inequalities draw a graph to show the region in which *p* and *q* must lie.

f) Allowing for depreciation and all other expenses the expected profit from each car of type *P* is £300 per annum and from *Q* £480 per annum. What is the total expected profit per annum?

g) Using your graph find what values of *p* and *q* give the maximum expected profit. (*Hint* Find the maximum expected profit at or near each node. *p* and *q* must be whole numbers.)

9   A firm produces two new types of toy, *X* and *Y*. Let the number produced per week be *x* of *X* and *y* of *Y*.

a) Each toy of type *X* requires 50 minutes of joiners' time, and *Y* requires 60 minutes. How much joiners' time is required altogether each week?

b) If the factory works not more than 50 hours a week and there are 10 joiners, how much joiners' time is available each week? Give your answer first in hours and then in minutes.

c) Write down an inequality expressing the results of *a* and *b*.

d) Each toy of type *X* requires 70 minutes of metal workers' time, and *Y* requires 30 minutes. What is the total metal workers' time required per week?

e) If the firm employs 7 metal workers, how much metal workers' time is available each week? Give your answer first in hours, and then in minutes.

f) Write down an inequality expressing the results of *d* and *e*.

g) If *Y* is very popular and the sales manager decides that at least 200 of *Y* must be produced each week, express this as an inequality.

h) Draw a graph to represent your three inequalities. Shade in lightly the area in which *x* and *y* must lie.

i) If the overall profit is £1 on each of *X* and 80p on each of *Y*, what is the total overall profit per week? Give your answer in pence.

j) What combination of *x* and *y* will make this overall profit a maximum? (*Hint* Find the overall profit at or near each node in turn. Give your answers to the nearest ten units.)

k) How many of *X* and how many of *Y* should be produced each week?

10   A confectionery manufacturer produces two types of boxes of chocolates, Expectation (*E*) and Realization (*R*).

a) If he produces *e* of *E* and *r* or *R* and the total daily output must exceed 800 boxes, express this as an inequality in *e* and *r*.

b) There must not be more than 500 of *R*. Express this as an inequality.

c) The number of *E* must not exceed the number of *R* by more than 300. Express this as an inequality.

d) Draw a graph to illustrate these three inequalities and show by shading the region in which *e* and *r* must lie.

e) If the profit on *E* is 20p a box and on *R* is 10p a box, what is the total daily profit?

f) Find from your graph the values of *e* and *r* which will make this profit a maximum. What is the maximum daily profit?

11   The manager of a large hostel wishes to re-carpet as many as possible of his single rooms. He decides to use Loomax and Easilay.

a) If he carpets *x* rooms with Loomax at a cost of £120 a room and *y* rooms with Easilay at £150 a room, what is the total cost of the carpet?

b) If he must not spend more than £3000 on carpet, write down an inequality concerning cost.

c) If it takes one man $\frac{1}{2}$ day to carpet a room with Loomax and $\frac{1}{3}$ day with Easilay find the time taken to carpet the rooms in *a*.

d) If he has 4 men available but the work must be finished in 10 days or less write an inequality concerning time. Multiply through by 6 to get rid of the fractions. (*Hint* In how many days could one man do the same work as 4 men in 10 days?)

e) If there are 100 rooms altogether, write down an inequality concerning the number of rooms. (*Hint* He can carpet less than 100, but not more than 100.)

f) Draw a graph showing the region in which *x* and *y* must lie.

g) From the graph find the maximum number of rooms he can carpet. (*Hint* Find the value of $x + y$ at or near each node.)

h) From your graph you can see that one of the inequalities was superfluous. Which?

If the life of Loomax is estimated at 12 years and the life of Easilay at 9 years and he decides to carpet in a way which gives maximum total life (where total life is $12x + 9y$ years) what combination of *x* and *y* should he now use and how many rooms can he now carpet?

12 A farmer uses two kinds of fertiliser, Phosco and Nutri. He buys this in bulk in 50 kg bags and stores it ready for use. The capacity of his store is 75 m^3. He would like to fertilise 100 hectares but he must not spend more than £3500 on fertilisers altogether. The two fertilisers are used separately on different fields. Here are further details:

	Storage room per bag	No. of bags used on one hectare	Cost per bag
Phosco	0.1 m^3	6	£6
Nutri	0.15 m^3	7	£5

a) If he orders *p* bags of Phosco and *n* bags of Nutri, write down an inequality concerning storage space. Multiply it by 20 to eliminate fractions and decimals.

b) Write down an inequality concerning cost.

c) Draw a graph and shade the area in which *p* and *n* must lie.

d) How many hectares can he fertilise *i*) with *p* bags of Phosco, *ii*) with *n* bags of Nutri, *iii*) altogether?

e) Find from your graph the combination of *p* and *n* (to the nearest 10 bags of each) which will allow him to fertilise the largest possible area.

f) How many of his 100 hectares can he fertilise?

13 A farmer has 10 fields in which he grazes sheep and bullocks. Each field is large enough for 50 bullocks or 200 sheep. He employs 4 men regularly and one man can look after 200 sheep or 200 bullocks. The profit per bullock is £50 and per sheep is £10. If he uses *b* fields for bullocks and *s* fields for sheep,

a) write down an inequality concerning the number of fields,

b) write down an inequality concerning the number of men. (*Hint* *b* fields of bullocks require 50*b*/200 men. How many men do *s* fields of sheep require?) Multiply this inequality by 4 to eliminate fractions.

c) Draw a graph and shade in lightly the region in which *b* and *s* must lie. Mark the grid points in this region, i.e. the points where both *b* and *s* have integral values.

d) What is the profit from *b* fields of bullocks, i.e. from 50*b* bullocks?

e) What is the profit from *s* fields of sheep? What is the total profit?

162

f) From your graph find the combination of *b* and *s* which will give the greatest profit. Read only the grid points. Fractions of fields are not allowed.

g) How many bullocks should he graze, and how many sheep?

14 A bus company has a fleet of 10 double deckers seating 65 and 2 single deckers seating 40. They are under contract to move 520 workmen between a factory and the nearby town twice daily. The double decker requires a crew of 2, one conductor and one driver. The single decker is 'one man operated'. There is a pool of 24 drivers all of whom are willing to act as conductors when required. If it costs twice as much to run a double decker as to run a single decker, how many of each should be used? (To answer this question consider *d* double deckers and *s* single deckers. Write down an inequality concerning *d*, another concerning *s*, one in *d* and *s* concerning drivers, and one in *d* and *s* concerning passengers. Cancel them down to the simplest form possible. Draw the graph, shade the region in which *d* and *s* must lie and find the selection that gives the minimum cost, i.e. the minimum value of $2d + s$.)

15 A manufacturer makes a mustard pickle in which two of the main ingredients are cauliflower and gherkin. For the best flavour he finds that in 100 kg of pickle there should be between 8 and 13 kg of cauliflower and between 9 and 13 kg of gherkin. The upper and lower limits are allowed, but the combined weight of cauliflower and gherkins must not exceed 20 kg.

a) If in 100 kg of pickle there are *c* kg of cauliflower and *g* kg of gherkin write down three inequalities expressing the above statements. (Two of these will be 'double' inequalities, i.e. of the form $a \leqslant x \leqslant b$.)

b) Draw a graph and show, by shading, the region in which *c* and *g* must lie.

c) If in a certain year the cost of 1 kg of gherkins is three times the cost of 1 kg of cauliflower, which mixture will give the least cost of 'cauliflower plus gherkin' and which mixture will give the greatest cost?

d) Answer question *c* for a year in which cauliflower costs twice as much per kilogram as gherkins.

16 The sum of two numbers is less than 30. The sum of the first plus twice the second is greater than 20. Three times the first plus twice the second is greater than 36. Both numbers are integral multiples of 5. (10 and 5 count as multiples of 5.)

a) Calling the numbers *x* and *y*, write down three inequalities in *x* and *y*.

b) Draw a graph and shade very lightly the area in which *x* and *y* must lie. Mark the lattice points with a heavy dot (i.e. points in the shaded area at which both *x* and *y* are multiples of 5).

c) List all the possible pairs of values of *x* and *y*.

d) How many lattice points are there in the shaded area?

e) If the signs of the three inequalities in *a* were changed from < and > to ≤ and ≥, mark the additional lattice points on your graph with a small cross. How many are there?

f) List the additional pairs of values.

17 Mrs Smock runs a 'factory' in her own home making clothes for teenagers. She has six fully automatic sewing machines and employs six local women as machinists and three others for hand operations and general duties. The working week is 30 hours. At a bankrupt sale she buys 600 metres of assorted cloth of which 300 metres only is suitable for dresses but all is suitable for blouses. A dress takes 2·5 m of material and 3 hours of machine time. A blouse takes 1·5 metres of material and 1 hour of machine time. Hand time for dresses and blouses is $\frac{1}{2}$ hour and $\frac{3}{4}$ hour respectively. She must get the work finished in three weeks.

a) How many hours' work can the 6 machinists do in 3 weeks?
b) How many hours' work can the 3 hand workers do in 3 weeks?
c) If she makes *d* dresses and *b* blouses, write down four inequalities in *b* and/or *d*. (Two will concern material and two will concern time.)
d) Draw a graph and shade the region in which *b* and *d* must lie.
e) If there is a profit of 75p on dresses and 50p on blouses, what combination of *b* and *d* will give the greatest profit?
f) Can you suggest another combination which would give a little less profit but might be better? Say why it would be better.

∗ 18 A firm marketing table sauces produces two main brands, Piquant and Puissant. Both contain two rare spices, Camoyard and Fennet, which are in short supply, only $1\frac{1}{2}$ kg of the former and 2 kg of the latter being available each week. The weights of each spice in mg per kg of sauce are shown below.

	Camoyard	Fennet
Piquant	40	100
Puissant	100	50

The sauce is sold in bottles each holding 200 g, and the bottles are packed in cases, 20 bottles to a case. Output is limited to 5500 cases a week of which at least 500 must be Puissant. If the firm produces *x* cases of Piquant and *y* cases of Puissant each week,

a) how many kg of sauce are there in each case?
b) how many grams of Camoyard are there *i*) in one case of Piquant, *ii*) in *x* cases of Piquant, *iii*) in *y* cases of Puissant?
c) how many grams of Fennet are there *i*) in 1 case of Piquant, *ii*) in *x* cases of Piquant, *iii*) in *y* cases of Puissant?
d) Using the results of *b* and *c* write down two inequalities about the supply of Camoyard and Fennet. Simplify them as much as possible.
e) Write down an inequality about the total number of cases produced each week.
f) Write down an inequality about the number of cases of Puissant.
g) Draw a graph and show the region in which *x* and *y* must lie.
h) If the net profit is £1 per case of Piquant and 80p per case of Puissant, what is the total net profit per week in pence (in terms of *x* and *y*)?
i) What combination of *x* and *y* will give the greatest overall profit?

∗ 19 In a certain university candidates applying for admission to read mathematics are asked to sit a test containing questions in modern mathematics, statistics and conventional mathematics (pure and applied). Part I of the paper consists of short questions and Part II of longer questions, both parts containing questions in all four disciplines. A single mark, however, is awarded for each part. These marks are 'processed', i.e. scaled by certain factors and then added together. The scale factors used by each of the four departments are shown below. If all three of a candidate's processed marks are equal to or greater than the figures shown in the last column, he is then invited for an interview.

Department	Scale factor Part I	Scale factor Part II	Minimum mark required
Modern Maths	2	1	80
Statistics	1	1	60
Conventional Maths (*Pure or Applied*)	3	10	300

164

a) Taking x and y as the actual marks in Parts I and II write down three inequalities in x and y.

b) Draw a graph and show by shading the lower bounds of the region in which x and y must lie if the candidate is to be invited for interview.

c) What is the minimum total mark before processing $(x+y)$ which a candidate must obtain to be invited for interview?

✻ 20 A sailing club based on a canal reservoir has 100 mooring bays each of which will take two sailing dinghies or one canal cruiser from the adjacent canal. Half the mooring accommodation is reserved for dinghies only and the other half may be used for either dinghies or cruisers. In the summer the club house is open daily from 10 a.m. to sunset. To avoid overcrowding the management try to keep the number of adults using the club house at any one time to 75. A sailing dinghy has an average crew of two adults who spend about one quarter of the open time in the club. A cruiser has an average crew of four adults who spend about one sixteenth of the open time in the club.

a) If there are d dinghies, how many mooring bays do they occupy? How many mooring bays do c cruisers occupy? Write an inequality about mooring bays.

b) Write an inequality about the total number of cruisers.

c) How many adults are there on c cruisers? How many on d dinghies? What is the average number of adults in the club at any one time? (*Hint*   Add one sixteenth of the first number to a quarter of the second.)

d) Write down an inequality about the number of adults in the club. Multiply by a suitable factor to get rid of the fractions.

e) Draw a graph and shade the region in which c and d must lie.

f) From your graph state the greatest number of cruisers that should be moored.

g) State also the greatest total number of boats that should be moored.

h) If the mooring charge for a cruiser is 1·5 times the charge for a dinghy, state the number of each that should be moored to get the maximum mooring fees.

✻ 21 A charitable organisation sends food parcels to prisoners all over the world. By international agreement these parcels must not exceed 16 units of volume and 10 units of weight. The organisation will not spend more than £8 on the contents of any one parcel. The parcels are made up entirely of Vit-paks and Pro-paks which are supplied direct by the manufacturers. The table gives details of these 'paks'.

	Units of volume per pak	Units of weight per pak	Cost (pence)	Units of vitamin per pak	Units of protein per pak
Vit-paks	2	1	40	6	4
Pro-paks	1	1	100	2	7

a) If each parcel contains v Vit-paks and p Pro-paks, write three inequalities expressing the information in the first three columns of the table. Cancel them down to their lowest terms.

b) Draw a graph and shade the region in which v and p must lie.

c) From your graph find the composition of the parcel most suited to prisoners suffering from vitamin deficiency (i.e. parcels for which $6v+2p$ is a maximum).

d) Find also the composition of parcels for prisoners suffering from protein deficiency (i.e. parcels for which $4v+7p$ is a maximum).

18 Everyday Arithmetic and Exploring More Byways

18A Everyday Arithmetic

Simple Interest

1 a) What is the simple interest on £100 for 1 year at 6%?
  b) What is the simple interest on £100 for 1 year at 12%?
  c) What is the simple interest on £100 for 1 year at 15%?
  d) What is the simple interest on £100 for 1 year at r%?

2 a) What is the simple interest on £300 for 1 year at 12%?
  b) What is the simple interest on £500 for 1 year at 12%?
  c) What is the simple interest on £800 for 1 year at 12%?
  d) What is the simple interest on £$(n \times 100)$ for 1 year at 12%?
  e) What is the simple interest on £P for 1 year at 12%?

3 a) What is the simple interest on £100 at 12% for 1 year?
  b) What is the simple interest on £100 at 12% for 3 years?
  c) What is the simple interest on £100 at 12% for 7 years?
  d) What is the simple interest on £200 at 12% for 7 years?
  e) What is the simple interest on £600 at 12% for 7 years?
  f) What is the simple interest on £600 at 8% for 7 years?
  g) What is the simple interest on £600 at 8% for 6 years?
  h) What is the simple interest on £400 at 10% for 5 years?
  i) What is the simple interest on £$(n \times 100)$ at r% for t years?
  j) What is the simple interest on £P at r% for t years?

4 Calculate the simple interest on the following. You are given the sum of money (called the *principal*), the rate per cent and the time.

  a) £500, 6%, 3 years   b) £800, 11%, 5 years   c) £900, 13%, 4 years
  d) £600, 8%, 8 years   e) £1250, 8%, 9 years

5 Calculate the simple interest on the following:

  a) £340, 4%, 2 years   b) £720, 6%, 4 years   c) £930, 7%, 3 years
  d) £630, 12%, 2 years   e) £225, 9%, 8 years

166

6 In everyday life, simple interest is required correct to the nearest penny. In the following examples logarithms or slide rule will not give enough accuracy, so work the answers longhand or use a calculator.

a) £360, $11\frac{1}{2}\%$, 3 years 9 months
b) £430, $7\frac{3}{4}\%$, 5 years 6 months
c) £290, $8\frac{1}{2}\%$, 8 years 3 months
d) £1020, $11\frac{1}{4}\%$, 7 years
e) £1210, 9%, 6 years 2 months

7 Using a calculator if possible, find the simple interest on the following. (One of these can be worked in your head.)

a) £243 at $7\frac{1}{4}\%$ for 7 months
b) £516 at $8\frac{3}{4}\%$ for 2 years 11 months
c) £903 at $5\frac{1}{2}\%$ for 1 year 8 months
d) £816 at 12·5% for 8 years
e) £732 at $13\frac{1}{4}\%$ for 4 years 4 months

8 After how many years will the total interest add up to a sum equal to the principal (or, loosely speaking, in how many years will the money invested be doubled):

a) at 5%   b) at 10%   c) at 8%   d) at $12\frac{1}{2}\%$ e) at 15% simple interest?

9 As we have already seen, the simple interest £I on a principal of £P for t years at $r\%$ is given by the formula $I = \dfrac{Ptr}{100}$. Change this formula to make a) P the subject, b) t the subject, c) r the subject.

10 Using the results of question 9 or otherwise, answer the following questions:

a) The interest on a sum of money for 5 years at 8% was £240.
What was the sum of money?
b) If the interest on a loan amounted to £230 for 3 years 6 months at 8%, what was the original loan? Answer to the nearest £1.
c) The interest on an investment was £54 for the first six months. If the rate per cent is 9%, how much was invested?

11 Using the results of question 9 or otherwise, answer the following questions:

a) If the interest on £400 in 3 years is £78, what is the rate per cent?
b) After how many years will the interest on £650 at 12% come to a total of £468?
c) At what rate per cent will the interest on £543 amount to £122.18 in three years?
d) £820 is invested at $8\frac{1}{2}\%$. After how long will the interest exceed £400? (Give your answer to the nearest month.)

Compound Interest

In compound interest, instead of the interest being paid to the lender every year (or more frequently by arrangement), it is added to the principal. Interest plus principal is called the *amount* and the next year's interest is calculated on the new amount.

12 a) If I lend £100 at 5% interest, what interest will it earn in the first year? What will be the amount at the end of the first year?
b) The interest for the 2nd year will be 5% of this amount. What will be the interest for the 2nd year? What will be the amount at the end of the 2nd year?
c) If you have calculated correctly so far, you will know the amount at the end of the second year is £110.25. Interest for the 3rd year will be 5% of this amount. You can see that the calculation soon gets very wearisome. So tables of compound interest are used. These tell you the amount after 1, 2, 3 years, etc., at various rates of interest.

Here is a typical table:

£100 at Compound Interest

Rate per cent	Amount after 1 year	Amount after 2 years	Amount after 3 years	Amount after 4 years
3%	£103.00	£106.09	£109.27	£112.55
4%	£104.00	£108.16	£112.49	£116.99
5%	£105.00	£110.25	£115.76	£121.55

13 Using the table in question 12, answer the following questions:

a) At 4% compound interest what does £100 amount to after 2 years?
b) At 3% compound interest what does £100 amount to after 3 years?
c) At 5% compound interest what does £100 amount to after 4 years?
d) At 5% compound interest what does £200 amount to after 4 years?
e) At 5% compound interest what does £500 amount to after 4 years?
f) At 5% compound interest what does £450 amount to after 4 years?
g) At 4% compound interest what does £300 amount to after 3 years?
h) At 3% compound interest what does £700 amount to after 2 years?

14 The table you have been using would not be accurate enough for large sums of money, so here is another table worked to 4 decimal places. This will give correct answers for sums of money up to £10 000 and the error will not be more than one penny for sums up to £100 000.

£100 at Compound Interest

	1 year	2 years	3 years	4 years	5 years	6 years	7 years
3%	£103.0000	£106.0900	£109.2727	£112.5509	£115.9274	£119.4052	£122.9874
4%	£104.0000	£108.1600	£112.4864	£116.9859	£121.6653	£126.5319	£131.5932
5%	£105.0000	£110.2500	£115.7625	£121.5506	£127.6282	£134.0096	£140.7101
6%	£106.0000	£112.3600	£119.1016	£126.2477	£133.8226	£141.8519	£150.3631
7%	£107.0000	£114.4900	£122.5043	£131.0796	£140.2552	£150.0730	£160.5782
8%	£108.0000	£116.6400	£125.9712	£136.0489	£146.9328	£158.6874	£171.3824
9%	£109.0000	£118.8100	£129.5029	£141.1582	£153.8624	£167.7100	£182.8039
10%	£110.0000	£121.0000	£133.1000	£146.4100	£161.0510	£177.1561	£194.8717

Using the table, answer the following questions:

a) What is the amount of £2000 at 7% compound interest for 5 years?
b) What is the amount of £2000 at 5% compound interest for 7 years?
c) What is the amount of £3000 at 4% compound interest for 6 years?
d) What is the amount of £800 at 8% compound interest for 3 years?
e) What is the amount of £7000 at 9% compound interest for 5 years?
f) What is the amount of £6500 at 6% compound interest for 4 years?
g) What is the amount of £3700 at 3% compound interest for 7 years?
h) What is the amount of £8150 at 10% compound interest for 4 years?

15 Peter's uncle dies and leaves Peter his entire fortune, well over £100 000. But there are difficulties in winding up the estate and as Peter needs more capital urgently for his own business, he borrows £10 000 from the bank at 8% compound interest. If it is three years before he receives an actual payment under the will, how much must he repay to the bank?

16 David, who has recently set up home, wants to borrow £10 000 for home improvements. He will be able to repay it in four years' time when he is 25 and receives payment of a legacy. He can have the money at 10% simple interest (this interest being paid quarterly), or at 9% compound interest.

a) If he chooses simple interest, how much interest does he pay each quarter?
b) How much interest does he pay altogether?
c) What is his total repayment?
d) If he chooses compound interest, what is his total repayment after four years?
e) With which method does he pay more? How much more?

17 What is the monthly interest payment (simple interest) on:

a) £360 at 8%   b) £400 at 5%   c) £600 at 10%   d) £300 at 7%
e) £700 at 5% (Answer to nearest 1p.)

18 A man buys a house for £10 000 and six years later sells it for £16 000.
a) What is his profit?
b) What would have been his profit if he had lent £10 000 at 10% simple interest for six years?
c) What would have been his profit (to the nearest £) if he had lent £10 000 at 8% compound interest for six years?
d) What would have been his profit (to the nearest £) if he had lent £10 000 at 10% compound interest for six years?

18B   More Everyday Arithmetic

Hire Purchase

1 Mrs Snow buys an automatic washing machine for £140. She pays £20 deposit. An interest charge of 14% is then added to the balance, and the total sum divided by 24 to find her monthly repayment for the next two years.

a) What is the total interest charge? The law requires that this should be entered on the agreement.
b) What is the total outstanding balance after addition of the interest charge?
c) If this is spread over two years, what is the monthly repayment?
d) What is the total sum she pays for her washing machine?

2 Harry buys a bicycle for £68. He pays £20 deposit. A service charge of 15% is added to the outstanding balance.

a) How much is the service charge?
b) What is the new outstanding balance?
c) If he pays this in 24 monthly instalments, how much is each instalment (to the nearest 10p)?

3 A secondhand car is bought for £800. The deposit is £200. 10% is added to the balance, and this is divided into 12 equal monthly payments.

a) What is the total interest paid?
b) What is the monthly payment?
c) What is the total price paid for the car?

4 A 3-piece suite is purchased for £830 and payment is spread over 3 years. If a deposit of £200 is paid and the service charge amounts to 20%, what will be the monthly repayment (rounded up to the nearest 10p)?

5 A motorcycle is bought for £240 and a deposit of £60 is paid. Payment of the balance is spread over 12 months and a credit charge of 11% is made. What is the monthly repayment? What is the total cost of the motorcycle?

6 In the following examples the monthly repayment is to be rounded up to the nearest 10p. This will give a total repayment slightly in excess of the sum due, so the last monthly repayment must be reduced. In each case state the monthly repayment, and the payment due in the last month.

a) Cost £440.   Deposit £120.   Spread over 36 months. Interest charge 20%.
b) Cost £92.   Deposit £24.   Spread over 12 months. Interest charge 9%.
c) Cost £880.   Deposit £300.   Spread over 24 months. Interest charge 14%.

7 A young man buys a television set for £120. He pays £30 down and can have the balance on hire purchase. A credit charge of 10% would be added, and he would pay in 12 equal monthly instalments. How much a month would he pay? How much would he pay altogether?

∗ 8 If you like long calculations, answer the following questions.
The young man in question 22 had a credit card with one of the major banks, and decided to use this to pay the balance of £90 on his television set. Interest is added at 2% each month to the total sum owing, and he can pay off at whatever rate he likes, subject to a certain minimum. He pays £15 a month for the first three months, £5 only in the 4th month and £10 a month thereafter, making the final payment in the 9th month. How much is this final payment? How much has it cost him altogether? Is this cheaper than buying on hire purchase?
To help you work out your answer, draw up a table like the one below and complete the remaining lines:

	Owing at beginning of month	Interest at 2% a month	Total	Payment	Balance c/f
1st month	£90.00	£1.80	£91.80	£15.00	£76.80
2nd month	£76.80	£1.54	£78.34	£15.00	£63.34
3rd month	£63.34				

Mortgages

9 A couple buy a house for £23 000 and pay £5000 down. They pay the balance over 20 years, the repayment being £158 monthly. How much do they pay for the house altogether?

10 Another couple buy a house for £14 000. They pay £3000 down and arrange a building society mortgage to cover the balance of the cost. If this is spread over 10 years the monthly repayment will be £132, but if it is spread over 20 years the monthly repayment will be £96.90.
How much will the house cost altogether if they spread the payment over *a)* ten years, *b)* twenty years?

11 Calculate the total price of the following houses:

a) Purchase price £16 000. Down payment £4000. Repayment over 15 years at £120 a month.
b) Purchase price £8500. Down payment £3500. Repayment over 10 years at £58 a month.
c) Purchase price £13 200. Down payment £10 000. Repayment over 5 years at £66 a month.

d) Purchase price £17 000. Down payment £1700. Repayment over 30 years at £118 a month.

٭ 12 A purchaser buys a house for £12 500 and pays £8500 down. He borrows the remaining £4000 from a friend who takes out a mortgage and charges him interest at the rate of 10%. The interest is calculated once a year as 10% of the sum owing at that time. So the first year's interest is £400. If the purchaser pays off his debt at the rate of £80 a month, how long will it take to clear the mortgage? How much should his last payment be? How much does he repay to his friend altogether (to nearest £)? How much does the house cost him altogether (to nearest £)? (*Hint* Make up a table similar to the one in question 8, but each line referring to one year instead of one month. In the last year charge interest for only ⅔ of a year, not a full year.)

13 Building societies naturally insist on a house on which they are lending money being fully insured. They often arrange the insurance themselves and charge the premium to the borrower. This means that the monthly repayment on the mortgage is increased.
If the rate of insurance is £1.25 per annum per £1000, what would be the cost of insurance per annum for the following houses?

   a) Insured for £20 000   *b*) Insured for £25 000   *c*) Insured for £15 000
   d) Insured for £12 500   *e*) Insured for £22 500

14 Building societies often insure a house for more than the purchaser actually pays for it (as the cost of rebuilding a house that was completely destroyed could well exceed the actual purchase price). If a house was insured by a building society for £28 000,

   a) what would be the annual premium at £1·25 per £1000 insured?
   b) what would be the addition to the monthly repayment?

15 Answer question 14 for a house that was insured for *a*) £30 000   *b*) £18 000.

16 A purchaser pays £22 500 for a house. He puts down £6500 and obtains the rest on a 20 year mortgage, the repayment rate without insurance being £8.81 a month on every £1000 lent. If the society insures the house for £29 000, the insurance rates being £1.25 per £1000 insured, what will be the monthly payment to the building society?

17 Repeat question 16 for a house purchased for £18 000, the deposit paid being £3000 and the insurance being for £23 000. The mortgage repayment rate is still £8.81 per month per £1000 borrowed.

18 Repeat question 16 for a house purchased for £12 500, the deposit paid being £3500 and the insurance being for £16 000. The mortgage repayment rate is still £8.81 per month per £1000 borrowed.

18C Exploring More Byways

1 How to create area
Using any suitable scale, draw accurately four right-angled triangles with sides 20 and 21 units. Using Pythagoras' theorem, the hypotenuse is 29 units. The area of each triangle is 210 square units. Check these calculations for yourself.

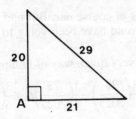

Now cut out these four triangles and fit them together to form a square of side 29 units with the vertices A at the centre. The area of this square is 29×29 which is 841 square units. But the area of the four triangles is only 840 square units. You have apparently created an area of 1 square unit. Explain the fallacy.

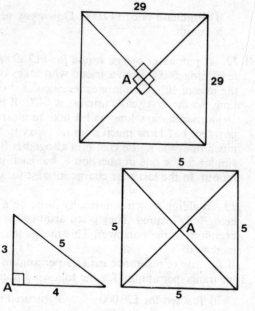

2 If you were unable to spot the fallacy in question 1, it may be easier to spot if you use four right-angled triangles each of sides 3 cm and 4 cm. In each triangle the hypotenuse is 5 cm, and the area is 6 cm². The four triangles fit together to form a square of side 5 cm. The area of this square is 25 cm², but the area of the four triangles is only 24 cm². Can you see the fallacy this time?

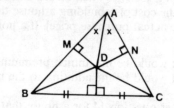

3 Here is a proof that every triangle is isosceles. Can you find the fallacy? ABC is *any* triangle. The bisector of angle A and the perpendicular bisector (or mediator) of BC meet at D. DM and DN are drawn perpendicular to AB and AC. Any point on the bisector of an angle is equidistant from the sides, so DM and DN are equal. Triangles ADM and ADN are both right angled, and as they have the same hypotenuse AD and equal sides DM and DN, from Pythagoras' theorem or from congruent triangles the third sides AM and AN must be equal.

Any point on the mediator of BC is equidistant from B and C. So DB must equal DC. Triangles BDM, CDN are both right angled. They have equal hypotenuses BD and CD, and equal sides DM and DN, so by Pythagoras' theorem or from congruent triangles their third sides must be equal. So $BM = CN$. Hence $BM + AM = CN + AN$, i.e. $BA = CA$ and the triangle is isosceles. Can you see the fallacy?

4 'As I was going to St Ives
   I met a man had seven wives.
   Each wife had seven sacks.
   Each sack had seven cats.
   Each cat had seven kits.
   Kits, cats, sacks and wives –
   How many were going to St Ives?'

The answer is of course *one*. But had the wording been 'I overtook a man ...' how many then would have been going to St Ives?

5 Here is a very quick way of finding an accurate value of $\sqrt{2}$:

$$\sqrt{2} = \frac{45}{32}\left(1 + \frac{1}{2}\left(\frac{23}{2025}\right) - \frac{1}{4}\left(\frac{23}{2025}\right)t_2 + \frac{3}{6}\left(\frac{23}{2025}\right)t_3 - \frac{5}{8}\left(\frac{23}{2025}\right)t_4 + \frac{7}{10}\left(\frac{23}{2025}\right)t_5 \text{ etc.}\right)$$

172

To get the third term (t_3) you multiply the 2nd term (t_2) by $\dfrac{1}{4} \times \dfrac{23}{2025}$.

If you work the first three terms, each correct to 7 decimal places, you get a value of $\sqrt{2}$ correct to 6 decimal places. Four terms (worked to 9 d.p.) give $\sqrt{2}$ to 8 decimal places, and five terms (worked to 11 d.p.) give $\sqrt{2}$ to 10 decimal places. See how far you can get.

6 Here is a series for $\sqrt{3}$ and another for $\sqrt{5}$. The first series gives an answer correct to 10 d.p. after working only five terms. The second does this after only four terms.

$$\sqrt{3} = \frac{7}{4}\left(1 - \frac{1}{2}\left(\frac{1}{49}\right) - \frac{1}{4}\left(\frac{1}{49}\right)t_2 - \frac{3}{6}\left(\frac{1}{49}\right)t_3 - \frac{5}{8}\left(\frac{1}{49}\right)t_4 - \frac{7}{10}\left(\frac{1}{49}\right)t_5 \cdots\right)$$

$$\sqrt{5} = \frac{29}{13}\left(1 + \frac{1}{2}\left(\frac{4}{841}\right) - \frac{1}{4}\left(\frac{4}{841}\right)t_2 + \frac{3}{6}\left(\frac{4}{841}\right)t_3 - \frac{5}{8}\left(\frac{4}{841}\right)t_4 + \frac{7}{10}\left(\frac{4}{841}\right)t_5 \cdots\right)$$

Work out as many terms as you can manage.

7 Suppose you wanted to find the sum of the series of numbers 1, 3, 5, ... 23. One way to do it would be to write the series down twice over, the second time being 'back to front' with the last number under the first, etc., as shown:

1	3	5	7	9	11	13	15	17	19	21	23
23	21	19	17	15	13	11	9	7	5	3	1

If you add the pairs of numbers, what do you notice?
How many pairs are there? What is the total sum of all the pairs?
What is the sum of the numbers in the first series?

8 From question 7 you can now see that if you add the first and last numbers of a series of this kind and multiply by the number of terms in the series, you get twice the sum of the series. Dividing by two gives the required sum. Find the sum of these series:

 a) 7, 11, 15 ... 47   b) 6, 7, 8 ... 41   c) 10, 21, 32 ... 120
 d) The first sixteen terms of the series 4, 7, 10 ...
 e) The odd integers between 10 and 30.

✴9 All the terms in the series in questions 7 and 8 were whole numbers, so their sum must also be a whole number. Yet the instructions for finding the sum contained the step 'divide by 2'. Explain why this will never give you a fraction.

10 The series in questions 7 and 8 are known as 'arithmetic progressions'. Each term is formed by adding a 'constant difference' to the preceding term. What is the constant difference in these series:

 a) 3, 4, 5, 6 ...   b) 5, 7, 9, 11 ...   c) 2, 7, 12, 17 ...   d) 10, 20, 30, 40 ...

11 Here is another type of series, where each term is obtained by multiplying the preceding term by a 'common ratio'. What is this ratio in each of the following?

 a) 1, 2, 4, 8, 16 ...   b) 1, 3, 9, 27 ...   c) 6, 36, 216 ...

12 The series in question 11 are called 'geometric progressions'. *Provided* they begin with 1, their sum is easily found.

 a) Find the sum of the series 1, 2, 4, 8, 16, 32.
 b) Write down the next term in the series. What do you notice?

c) Check your findings by adding up the first 3, 5, 7 terms of the series and seeing if a similar result holds.

13 You should know by now that the sum of the series $1, 2, 4, 8, 16 \ldots$ etc., is one less than the next term of the series.
A similar, but slightly different, result holds for other geometric progressions beginning with 1.

a) What is the sum of $1, 3, 9, 27, 81$?
b) What is the next term in the series? Subtract 1 from it and see how it compares with the sum you found in a.
c) Check your findings for the first 3, 4, and 6 terms of the series.
d) Try and find similar (but slightly different) rules for the series $1, 4, 16, 64 \ldots$ and $1, 5, 25, 125 \ldots$

*14 In an earlier book we described a method used by French peasants in earlier times for multiplying two numbers between 5 and 10 on their fingers. To multiply 7 by 9, 7 is 2 greater than 5 so hold two fingers up on one hand. 9 is 4 greater than 5, so hold 4 fingers up on the other hand. Add the fingers that are up. This gives 6 (the ten digit of the product). Multiply the fingers that are down (3×1). This gives 3 (the units digit of the product). So the number is 63.
Note that thumbs count as fingers, and that a common sense adjustment has to be made when multiplying 6 by 6 or 6 by 7.
You should now be able to follow an algebraic proof of this method.

a) Call the numbers x and y. There are $x - 5$ fingers up on one hand. How many are up on the other hand? Add these. What do you get? This is the tens digit, so to get its real value multiply by 10. What does this give?
b) Now think of the fingers that are down. On one hand there are $(x - 5)$ up, so there are $5 - (x - 5)$ which is $10 - x$ down. How many are down on the other hand? These have to be multiplied $(10 - x)(10 - y)$. There are four terms in the product. Write it out in full. This is the units digit.
c) Now add the numbers you calculated in a and b and this gives the final answer. Cancel any terms that will cancel and you should be left with xy. This proves the method.

15 A number of 2 digits, x and y, is $10x + y$. The sum of its digits is $x + y$.

a) What is a number of 3 digits (x, y and z)? What is the sum of its digits?
b) What is a number of 4 digits, w, x, y and z? What is the sum of its digits?
c) Writing a 2-digit number as $9x + (x + y)$, $9x$ is divisible by 3, so if the sum of the digits $(x + y)$ is divisible by 3, the whole number is divisible by 3. Does a similar result hold for divisibility by 9?
d) In a similar way prove that for a 3-digit number if the sum of the digits is divisible by 3, the whole number is divisible by 3.
e) Repeat d for a 4-digit number.
f) Prove that if the sum of the digits for a 4-digit number is divisible by 9, then the number itself is divisible by 9.

*16 Divide a 3-figure number $100x + 10y + z$ by 11. The remainder is $x + 10y + z$. But as $10y = -y \pmod{11}$ we can write this as $x - y + z$. So if $x - y + z$ is divisible by 11, the whole number is divisible by 11.

a) Prove in a similar way that a 4-figure number with digits w, x, y, z is divisible by 11 if $-w + x - y + z$ is divisible by 11.
b) Prove a similar property for a 5-figure number and a 6-figure number.

Note Sometimes the expressions $x - y + z$, etc., will be zero, sometimes positive, sometimes negative. But the sign does not matter so we can 'add the first, third, fifth digits, etc., and subtract the 2nd, 4th, 6th digits, etc.' If the number so obtained is 0 or divisible by 11, the whole number is divisible by 11.

c) Test your rule on 132 573, 637 181 and 637 186.

* 17 The test for divisibility by 7 or 13 for a 6-figure number is to write the first 3 figures under the second 3 and subtract (do it the other way up where necessary to avoid negatives). If the result of the subtraction is divisible by 7 or 13, so is the original number. This is easily proved as follows. If the digits are a,b,c,x,y,z, the number is $100\,000a + 10\,000b + 1000c + 100x + 10y + z$ which can also be written $(100\,100a + 10\,010b + 1001c) + (100x + 10y + z) - (100a + 10b + c)$. The numbers in the last two brackets are the numbers we subtracted in our test. Can you prove the test is true?

Note The test works for a 4- or 5-figure number just as well. Add 0's (at either end) to bring it up to a 6-figure number.

* 18 In how many years does a sum of money double at 5% compound interest? To answer this question look at the following table which shows the 'amount' of £1 after 1,2,3 years, etc.

1 year	*2 years*	*3 years*	....	*7 years*
£1.05	£1.1025	£1.115 762 5		£1.407 101

To go from one column to the next you multiply by 1·05, so the figure in the column for the seventh year is $(1·05)^7$. Squaring this gives $(1·05)^{14}$ and the number is 1·9799. The next term, the amount for 15 years, is 2·0788. Clearly the amount has doubled in just over 14 years (nearer 14 than 15).
Now find in how many years a sum of money doubles

a) at 7% compound interest   *b*) at 9%   *c*) at 10%.

*19 Question 18 can be answered more easily using logarithms. The logarithm of 2 is 0·3010. The logarithm of 1·05 is 0·0212. Divide the first by the second and the answer is $\dfrac{0·3010}{0·0212} = 14·2$, i.e. the amount doubles in just over 14 years.

Using this method, find in how many years a sum of money doubles at 8% and at 11%. Find also at what rate per cent it doubles in five years.

*20 If instead of finding the amount at 5% in 1 year you found it at $2\frac{1}{2}$% for each of 2 periods of half a year, your answer would be $(1·025)^2$ and this is a little greater than 1·05. Find its exact value.
If you worked it for five-fifths of a year, each fifth at 1%, your answer would be $(1·01)^5$. What does this come to?
If you worked it for 10 periods each $\frac{1}{10}$ of a year, at 0·5% for each period, your answer would be $(1·005)^{10}$. (If you want to work this out, find by repeated squaring $(1·005)^2$, $(1·005)^4$ and $(1·005)^8$. Your answer will be the first of these 3 multiplied by the last.)
The amount does not increase indefinitely. It approaches a 'limiting value' of £1.0514 (to 4 d.p.).

18D Number Polygons

1 Here is a number rectangle.
It is made as follows:

2	3	4	6	7	7
2					1
1					5
3	3	2	7	6	5

A set of clues is given, the answers to which are numbers.

175

The last digit of any number is the same as the first digit of the next number. The numbers are written in a row from left to right but the 'linking digits' are written once only.

Here are the numbers which gave the above rectangle:
23, 346, 67, 771, 155, 567, 72, 2331, 122.

Whenever a digit is repeated, after writing the digit twice turn through 90° clockwise to write the next digit.

The last digit of the last number should match the first digit of the first number.

Here is another set of numbers. Write them as a rectangle.
267, 75, 554, 43, 3662, 27, 711, 14, 4322.

2 Draw this rectangle. Here are the clues:

a) $5 \times 5 \times 5$ b) 13×4 c) $2 \times 3^2 \times 4^2$

d) a^2 when $a = -9$ e) 14_{10} in base 3   f) $1638 \div 7$

g) 31×161 h) $(2 \quad 5) \begin{pmatrix} -2 \\ 3 \end{pmatrix}$

3 Here is another rectangle:

a) The fifth row of Pascal's triangle.
b) The angle of a regular octagon.
c) Value of $3x^2 + 5$ when $x = -4$.
d) $(2 \quad 0 \quad -1) \begin{pmatrix} 4 & 1 & 1 \\ 5 & 2 & 3 \\ 5 & -1 & 1 \end{pmatrix}$
e) 43_{10} written in base 2.
f) Volume of a cone in cm³ where height $= 14$ cm, radius $= 3$ cm (take π as $3\frac{1}{7}$).

g) 19% of £15 (answer in pence).
h) 4 consecutive integers with median 6·5.
i) Smaller share when £242 is divided in the ratio 4:7.
j) $116_9 - 18_9$ (answer in base 9).
k) Total of 3 numbers whose mean is 26.
l) 1, 1, 2, 3, 5, –, –, (next 2 numbers written as 1 number).
m) $31 \times \underline{\quad} = 9641$.

4 In this puzzle, the first answer is written across the page, as before, but the answers following are written either clockwise or anticlockwise, starting with the first or last digit of a previous answer. It is for you to decide which, e.g. if the first answer was 125, the second would start with either a '5' or a '1'. It might be 568, giving 12568, or be 107 giving 70125. When all the clues are solved, the completed set of solutions should link up to form a polygon. (The rule about turning corners still holds.)

Clues

a) Solution of $4x + y = 52$, $x - y = 8$ written x, y.
b) If $f : x \rightarrow (3x + 1)^2$, find $f(4)$.
c) $(1 \quad 4 \quad -5) \begin{pmatrix} 2 & -4 & 3 \\ 3 & 0 & 2 \\ 1 & -1 & 2 \end{pmatrix}$
d) Find the image of (5,2) under the transformation represented by $\begin{pmatrix} 0 & 2 \\ -1 & 5 \end{pmatrix}$.
e) In fig. 1 find the shaded area in cm² (take π as $3\frac{1}{7}$).
f) Sum of the prime factors of 705.
g) Angle BAD in fig. 2.
h) $\dfrac{3 \cdot 08 \times 10^7 \times 4 \times 10^8}{1 \cdot 6 \times 10^{14}}$
i) A point in the region $x + y \leqslant 28$, $y \leqslant 3x$, $y \geqslant 10$ such that y is as large as possible.
j) Volume in cm³ of the solid shown in fig. 3.
k) If $A = (2 \quad 3)$ and $B = (-1 \quad -6)$ find $7A + 2B$.
l) £1000 reduced by $5\frac{1}{2}$%.

Fig.1

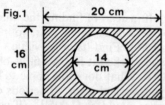

Fig.2

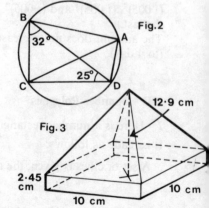

Fig.3

5 Draw a tessellation of equilateral triangles. Solve the clues, writing the first answer across the grid in the usual way. After a repeated digit, turn and continue in a new direction until another repeated digit occurs. (In all cases cross a line, not a point, when placing the next digit.) Keep adding numbers in the same sense and the final answer should link up with the first.

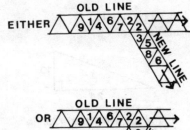

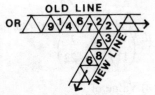

Clues

a) Three consecutive integers with a total of 18.

b) $2x^2$ when $x = 6$.

c) 47_{10} written in base 4.

d) 15×23.

e) $\begin{pmatrix} 2 \\ 8 \\ 3 \end{pmatrix} + \begin{pmatrix} 3 \\ -1 \\ 5 \end{pmatrix}$

f) If $x + y = 19$ and $x - y = 3$, find the value of xy.

g) $4.87\,\text{m} + 3.36\,\text{m}$, answer in cm.

h) First two primes after 35.

i) $459 \div 27$.

j) abc^2 where $a = 3$, $b = 5$, $c = -7$.

k) $1000 - 408$.

l) £200 increased by $5\frac{1}{2}\%$.

m) 29×6.

n) £112 divided in the ratio 3:4.

o) $2^6 - 20$.

p) $\dfrac{(9 \times 10^4)^2}{2 \times 10^7}$

In numbers 6 to 12 use an equilateral triangle tessellation. Turn after a second repeated digit. (See also number 5.)

6 a) $5 \times 12 + 1$.

b) 4 numbers with mean 5, mode 5, first one 8 less than second.

c) $\begin{pmatrix} 2 & 3 & 1 \\ -1 & 1 & 1 \\ 6 & 2 & 3 \end{pmatrix} \begin{pmatrix} 1 \\ 1 \\ 0 \end{pmatrix}$

d) £80 increased by 10%.

e) 6 m, 260 cm, 34 mm; answer in mm.

f) £54 divided in the ratio 5:1.

g) $a^2 + ab$ when $a = -7$, $b = -6$.

h) $\sqrt{256}$.

i) Find x.

j) 25×27.

k) $x + y = 16$, $2x - 3y = 7$, find value of xy.

l) Volume in cm^3 of a cuboid 16 cm by 9 cm by 4 cm.

m) 414_{10} written in base 8.

7 a) $a^2 - 2a + 10$ where $a = 16$.

b) Circumference of circle of radius 7 cm $(\pi = \frac{22}{7})$.

c) $\begin{pmatrix} 3 \\ 14 \\ 7 \end{pmatrix} - \begin{pmatrix} -1 \\ 8 \\ 5 \end{pmatrix}$

d) 10110_2 written in base 10.

e) The sixth triangle number.

f) $21 - 4 \times 2$.

g) 11 dozen added to 10 score.

h) Value of x^2.

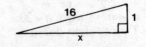

i) 65_7 written in base 8.

j) Value of x in metres.

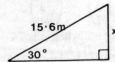

k) (x, y) if $x + y = 16$ and $y - x = 0$.

l) $(a^2 - a + 22)$ if $a = 30$.

m) $(1 \quad 2 \quad 3) \begin{pmatrix} 14 \\ 1 \\ 2 \end{pmatrix}$

8 a) $10000 - 1034$.
b) Value of x^2.

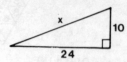

c) $a^2 + a + 4$ when $a = 25$.
d) $\begin{pmatrix} 11 \\ 11 \end{pmatrix} - \begin{pmatrix} 7 \\ -22 \end{pmatrix}$.

9 a) Value of x.

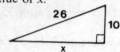

b) $\begin{pmatrix} 1 \\ 0 \\ 1 \\ 5 \end{pmatrix} + \begin{pmatrix} 3 \\ 1 \\ 1 \\ -2 \end{pmatrix}$

c) The first prime number (excluding 1) + eleventh prime number.
d) 3 consecutive integers forming a number divisible by 5.
e) $1000 - 422$.

10 a) $12 \times 11 + 1$.
b) $(15 \quad 9) \begin{pmatrix} 20 \\ 5 \end{pmatrix}$.
c) The tenth triangle number.
d) $24^2 + 10$.
e) 111111_2 in base 10.

e) $6 \times 50 + 1$.
f) Highest common factor of 24, 60, 108.
g) 14_{10} written in base 5.
h) $9(6x - 1)$ when $x = 9$.
i) 114_8 written in base 10.
j) 3 consecutive integers with mean 7.

f) 1011000_2 written in base 10.
g) $a^2 + 8$ when $a = 9$.
h) $(3 \quad 4) \begin{pmatrix} 12 \\ 15 \end{pmatrix}$.
i) The square of 25.
j) It has 2 digits and its factors are 2 prime numbers differing by 14.
k) H.C.F. of 1221 and 1331.
l) a,b,c,d if $(a \ b \ c \ d) - (3 \ -3 \ 0 \ 4) = (-2 \ 11 \ 1 \ -2)$.
m) 97.6% of £250.
n) Value of xy if $x + y = 13$ and $2x - y = 5$.

f) Value of x.

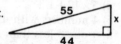

g) $1000 - 655$.
h) (x, y) if $x + y = 63$ and $y = \sqrt{64}$.
i) $9x^2$ where $x = -3$.

11 Enter the answers on a tessellation of equilateral triangles. The first answer is written across the page from left to right, and subsequent answers are added to either end of the set of numbers already entered. After a repeated digit, change direction.

a) 46_9 written in base 10.
b) $1000 - 745$.
c) $a^2 + a + 4$ when $a = 6$.
d) $\begin{pmatrix} 5 \\ 6 \\ 7 \end{pmatrix} - \begin{pmatrix} -1 \\ 3 \\ 4 \end{pmatrix}$.
e) $a^2 - 2$ when a is 24.
f) $(4 \quad 2) \begin{pmatrix} 8 \\ 1 \end{pmatrix}$.

g) 33_{10} written in base 7.
h) 36×12.
i) $\frac{1}{4}a^2$ when $a = 10$.
j) Smallest number with factors 7, 8, 28.
k) (x, y) if $y - x = 93$, $x + y = 105$.
l) $2 \times 300 - 11$.

12 (Same instructions as in 11.)
a) $(2 \quad 3 \quad 4) + (4 \quad 0 \quad -2)$.
b) x^2 when $x = -5$.
c) $1000 - 423$.
d) $(2 \quad 2) \begin{pmatrix} 30 \\ 7 \end{pmatrix}$.
e) 100_8 written in base 10.

f) Perimeter of figure in cm ($\pi = \frac{22}{7}$).

3.5 cm
14 cm

g) $a^2 + a - 8$ when $a = 20$.
h) Twenty-eight dozen.

i) Value of x^2.

j) 3 integers, mode 3, sum 8, smallest first.

k) Value of x in mm.

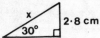

l) $18^2 + 1$.

m) Number less than 60. Sum of digits 11. Product of digits 30.

n) $154_7 + 455_7$ in base 7.

o) $6 + 2 \times 30$.

p) First multiple of 7 greater than 620.

q) $\frac{1}{2}(5a + 12)$ when $a = 6$.

r) $\begin{pmatrix} 4 \\ 1 \\ 3 \end{pmatrix} - \begin{pmatrix} 1 \\ 0 \\ -1 \end{pmatrix}$.

s) Curved surface area of a cylinder of diameter 7 cm, height 7 cm (take π as $\frac{22}{7}$).

Miscellaneous Examples C

Note This section contains a large number of C.S.E. and some 'O' level type questions.

C1

Say whether the following statements are true or false. When they are false, explain why and give a correct statement.

1 0.3 written as a fraction is $\frac{1}{3}$.

2 The solution of the equation $7-(x+6)=3(x-1)$ is $x=1$.

3 231_5 is even in base 10.

4 An area of $1\,500\,000$ cm^2 is equal to $15\,000$ m^2.

5 $\sqrt{0.8}=0.283$.

6 The median of the set of numbers $\{2 \quad 5 \quad 7 \quad 8 \quad 3\}$ is equal to the mean.

7 The total surface area of a solid cuboid 6 cm by 5 cm by 4 cm is 148 cm^2.

8 The image of (2,3) after reflection in $x=5$ is (8,3).

9 $\left(\dfrac{x}{3}-\dfrac{x}{6}\right)\div\dfrac{x}{2}$ is $\dfrac{x^2}{12}$.

10 An article costing £12 was reduced by 15% in a sale. It was not sold, so after the sale its price was raised by 10% of the sale price. It then cost more than its original price of £12.

11 In the figure, if $AE=12$ cm then $BD=6$ cm.

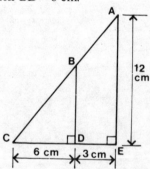

12 If $x-1=\sqrt{4y}$ then $y=\dfrac{(x-1)^2}{4}$.

13 If $a=6$, $b=-3$, $c=0$, $d=5$ then $ab<cd$.

14 The gradient of the line joining (1,2) to (5,−2) is −1.

15 The solution of the equation $2x^2-11x+5=0$ is $x=\frac{1}{2}$ or 5.

180

C2

Choose the correct answer from those given. Where appropriate, answers are given to 3 s.f.

1 A train travels at a steady speed of 96 km/h for 16 minutes. It covers a distance of:

  a) 6 km   *b)* 25·6 km   *c)* 256 km   *d)* 360 km   *e)* none of these.

2 If $A=\begin{pmatrix} 0 & 2 \\ -2 & 1 \end{pmatrix}$ the value of A^2 is:

  a) $\begin{pmatrix} 0 & 4 \\ 4 & 1 \end{pmatrix}$ *b)* $\begin{pmatrix} -4 & 4 \\ -4 & -5 \end{pmatrix}$ *c)* $\begin{pmatrix} -4 & 2 \\ -2 & -5 \end{pmatrix}$ *d)* $\begin{pmatrix} -4 & 2 \\ 2 & -3 \end{pmatrix}$ *e)* $\begin{pmatrix} -4 & 2 \\ -2 & -3 \end{pmatrix}$

3 If $y=\dfrac{4a+5b}{3}$ then a written in terms of y and b is:

  a) $\dfrac{3(y-5b)}{4}$ *b)* $\dfrac{15by}{4}$ *c)* $\dfrac{3y-5b}{4}$ *d)* $\dfrac{4y}{3}-5b$ *e)* none of these.

4 Simplify $\dfrac{1}{y}+\dfrac{5y}{4}$

  a) $\dfrac{1+5y}{y+4}$ *b)* $\dfrac{4+5y}{4}$ *c)* $1+5y$ *d)* $\dfrac{1+5y}{4y}$ *e)* $\dfrac{4+5y^2}{4y}$.

5 Factorise $x^2+3x-54$

  a) $(x-6)(x-9)$ *b)* $(x-6)(x+9)$ *c)* $(x+6)(x-9)$ *d)* $(x+6)(x+9)$
  e) none of these.

6 The perimeter of a 45° sector cut from a circle of radius 3·5 cm is:

  a) 5·5 cm   *b)* 9·65 cm   *c)* 11 cm   *d)* 9·75 cm   *e)* 54 cm.

7 The next prime number after 110 is:

  a) 111   *b)* 113   *c)* 117   *d)* 119   *e)* 121.

8 The interior angle of a regular polygon is 150°. The number of sides is:

  a) 8   *b)* 12   *c)* 21   *d)* 30   *e)* none of these.

9 In the figure if $AB=7$ cm, $BC=3$ cm, then the length of AC is:

  a) 10 cm   *b)* 3·16 cm   *c)* 7·48 cm   *d)* 7·62 cm   *e)* 58 cm.

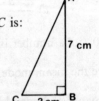

10 The line $3y=5-2x$ passes through the point:

  a) $(3,-2)$ *b)* $(2,-3)$ *c)* $(2,3)$ *d)* $(0,2·5)$ *e)* $(-2,3)$.

11 If A and B are sets such that $A'\cap B\neq\phi$, then it follows that:

  a) $A\subset B'$ *b)* $A\cap B\neq\phi$ *c)* $A\cup B\subset B$ *d)* $A'\neq\phi$ *e)* $B\subset A$.

12 If $\dfrac{4\cdot8\times10^3\times6\times10^5}{7\cdot2\times10^{-4}}=4\cdot0\times10^n$ the value of n is:

  a) 3   *b)* 4   *c)* 5   *d)* 12   *e)* 11   *f)* 13.

13 The value of $\dfrac{14\cdot5\times7\cdot66}{0\cdot045}$ to 2 s.f. is:

a) $0\cdot025$ b) $0\cdot25^{\cdot}$ c) 25 d) 250 e) 2500 f) none of these.

14 The value of $1\frac{2}{3}+\frac{4}{5}\times2\frac{1}{4}$ is:

a) $2\frac{8}{15}$ b) 3 c) $3\frac{7}{15}$ d) $5\frac{11}{20}$ e) none of these.

15 136_8 written in base 4 is:

a) 34 b) 46 c) 94 d) 1132 e) 232.

16 In the figure, $AB=BC$, $\angle BAC=64°$ and AB is parallel to CE. $\angle ECD$ is:

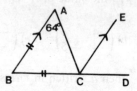

a) $26°$ b) $52°$ c) $58°$ d) $64°$ e) $116°$.

17 The image of the point $(3,1)$ after a $90°$ rotation anticlockwise about $(0,1)$ is:

a) $(0,3)$ b) $(0,4)$ c) $(0,-3)$ d) $(0,-2)$ e) $(2,-2)$.

18 The mean of the numbers $3,5,6,6,8,10,11$ is:

a) 6 b) 7 c) $6\cdot5$ d) 49 e) none of these.

19 The smallest share when £36 is divided in the ratio $2:3:4$ is:

a) £4 b) £8 c) £9 d) £12 e) £16.

20 In a furniture store all goods are increased in price by 18%. A carpet is now priced at £236. Before the increase it cost:

a) £131 b) £193.52 c) £200 d) £218 e) £223.

C3

1 Simplify a) $\frac{2}{3}+\frac{5}{6}$ b) $3-1\frac{4}{5}$ c) $2\frac{1}{2}\times1\frac{1}{5}$

d) $\dfrac{a}{3}-\dfrac{a}{4}$ e) $\dfrac{1}{a}\div\dfrac{3}{a}$ f) $\dfrac{a}{b}\times\dfrac{b}{a}$.

2 If 20% of a number is 280, what is the number?

3 Give the mean, mode and median of the numbers 1 2 2 3 5 7 7 7 8 8.

4 Copy out and complete these calculations which are all in the same base.

a) $\begin{array}{r} 123 \\ +\ 42 \\ \hline 2*0 \\ \hline \end{array}$ b) $\begin{array}{r} 123 \\ -42 \\ \hline *1 \\ \hline \end{array}$ c) $\begin{array}{r} 123 \\ \times\ 42 \\ \hline **1 \\ 1**20 \\ \hline ****1 \\ \hline \end{array}$

5 If $A'\subset B'$, draw a diagram showing the universal set and the two subsets A and B.

182

6 If $2x - 3 > 0$, what can you say about the values of x?

7 Give the next two terms of each of these sequences:

a) 1   4   9   16   25...
b) 1   2   3   5   8...
c) 23 29 31 37 41 43...

8 If $a * b$ means $a^2 - b^2$ find $5 * 3$ and $3 * 5$.
What is the value of $2 * (4 * 3)$?
If $a * b = b * a$, find the relationship between a and b.

C4

1 Goods are reduced in a sale by 35%. What is the sale price of a book originally priced at £5?

2 Given that $\sqrt{8} = 2 \cdot 83$ and $\sqrt{80} = 8 \cdot 94$, write down the values of

a) $\sqrt{800}$ b) $\sqrt{8000}$ c) $\sqrt{0 \cdot 8}$ d) $\sqrt{0 \cdot 08}$.

3 a) If $p = 4$, $q = 9$ and $r = 13$ find $\sqrt{pq + r}$.

    b) If $x = 3$, $y = 5$ and $z = -10$, find the values of $2x^2$, $x + y + z$, xyz, $\dfrac{x^2 y}{z}$, $\dfrac{y}{z}$.

4 Simplify   a) $4 \times 10^7 \times 8 \times 10^3$ b) $\dfrac{1 \cdot 4 \times 10^8}{2 \times 10^2}$ c) $(5 \times 10^{-3})^2$, giving your answers in standard form.

5 If £1 is equivalent to 455 Japanese yen, calculate the number of yen equivalent to £15.   What is the value of 800 yen in sterling to the nearest penny?

6 Sketch two quadrilaterals, one with line symmetry but no rotational symmetry, and the other with rotational symmetry but no lines of symmetry. Mark the lines of symmetry and the centre of rotation.

7 In the diagram, if $CD = 4$ cm find the lengths of AD, BD and AB, giving each correct to 2 s.f.

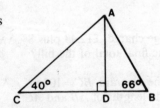

C5

1 Write down the direct route matrix for the network shown.

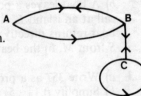

2 Find the value of   a) $49 \cdot 7 + 29 \cdot 3 + 50 \cdot 3 - 19 \cdot 3$
                          b) $39 \times 27 - 39 \times 17$
                          c) $8 \cdot 7^2 - 7 \cdot 7^2$.

3 If $x=1\frac{1}{2}$, $y=2\frac{2}{3}$ and $z=2\frac{1}{4}$, find the values of:

a) $x+y+z$ b) xyz c) $x+yz$ d) $2z\div(y-x)$.

4 Construct accurately the triangle ABC in which $AB=6$ cm, $BC=7$ cm and $AC=8$ cm. Draw the perpendicular from B on to AC. If it cuts AC at D, measure BD. Find the area of the triangle ABC. Show by shading the set of points P inside the triangle for which $PB \leqslant PA$.

5 Simplify   a) $a^3 \times a^2 \times a$ b) $(a^3)^2$ c) $a^3b^2 \times b^3a^2$ d) $a^3b^2c \div abc$

e) $\left(\dfrac{a}{b}\right)^2 \times ab$ f) $\sqrt{\dfrac{a^6b^4}{c^2}}$

6 $OACB$ is a parallelogram in which $\overline{OA}=\begin{pmatrix}1\\5\end{pmatrix}$ and $\overline{OB}=\begin{pmatrix}4\\2\end{pmatrix}$. Find $\overline{AB}$, $\overline{BA}$ and $\overline{OC}$.

7 Solve the following equations:

a) $4-x=1+x$ b) $5-3(2x-1)=2(5x+6)$ c) $\dfrac{3x}{5}-\dfrac{x}{3}=1\frac{1}{3}$

d) $\dfrac{x-1}{2}-\dfrac{2x+3}{6}=1$ e) $4^x=64$.

8 The diagram shows the end of a lean-to shed which is 3·5 m long. Calculate

a) the length of the sloping edge of the roof,
b) the area of the roof,
c) the area of the three sides and base (no back),
d) the capacity of the shed in m³.

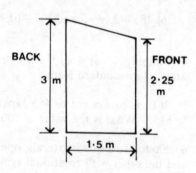

C6

1 A garage charges £12.43 plus 8% VAT for repairs to a car. What is the final total of the bill?

2 a) In the diagram, $BC=10$ km, $\angle A=90°$ and $\angle C=30°$. Find the lengths of AB and AC.

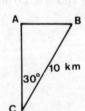

b) A boat leaves a port M on the mainland and sails 10 km on a bearing 030° to call at an island L. It then sails 4 km to an island S which is due east of L. The boat returns directly to M. Use your answers to a to find i) the direct distance of S from M, ii) the bearing of M from S (to the nearest degree).

3 a) Write 357 as a product of prime numbers.
b) Simplify i) $1\frac{7}{8}\times5\frac{1}{3}$ ii) $2\frac{1}{4}\div3\frac{3}{5}$.

4 Find the mean of the numbers 3, 5, 6, 8, 10.
Write down the mean of a) 33, 35, 36, 38, 40   b) 30, 50, 60, 80, 100.

5 If £1 is equivalent to 3·72 Swiss francs, find the value of £4.50 in Swiss francs. Find also the equivalent of 4·50 Swiss francs in sterling.

6 Find the length of *BD* in the diagram and hence the length of *BC*. Give your answers correct to 2 s.f.

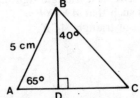

7 *ABC* is a triangle and *XY* is drawn parallel to *BC*, *X* and *Y* being points in *AB* and *AC* respectively. If *X* divides *AB* in the ratio *AX*:*XB*=1:2 and *AY*=3 cm, find the length of *AC*. If *BC*=6 cm, how long is *XY*?

8 The diagram shows two squares, one of side 4 cm and the other of side 2 cm. *AB*=*CD*=1 cm.
Draw the diagram accurately and construct two possible positions of the centre of an enlargement which maps the smaller square on to the larger square. Give the scale factor in each case.

C7

1 a) Find the exact answer when 430·95 is divided by 0·17.
b) Find 15% of £2.40.

2 A car maintains a speed of 88 km/h for 30 minutes along a motorway and then travels at 80 km/h for 15 minutes. Find to the nearest unit the average speed for the whole 45-minute journey.

3 What is the sum of the angles of a regular pentagon?
If a pentagon has two angles of 90° and three angles of *x*°, find *x*.

4 The ratio of teachers to pupils in a school is 1:19. If the school has 36 teachers, how many pupils are there? If the ratio is improved to 1:18, how many extra teachers are needed?

5 a) Calculate the height of an equilateral triangle of side 2 cm. Use your answer to find the volume of the triangular prism shown in fig. 1.
b) Fig. 2 shows the net of the prism being cut out of a rectangle of card. Find the dimensions of this rectangle and the area of card wasted.

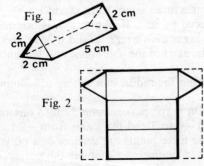

6 $\mathscr{E}$ ={3,4,5,6,7,8,9,10,11} and sets *A*, *B* and *C* are defined as follows:

$$A=\{x:x^2 > 30\} \qquad B=\{x:x \text{ is a factor of } 144\} \qquad C=\left\{x:\frac{2x+1}{3} \text{ is an integer}\right\}.$$

List *A*, *B* and *C* in full and draw a Venn diagram showing these sets.
Use the letters *A*, *B* and *C* to describe a sub-set of $\mathscr{E}$ which is empty.

185

7 If $p=101101_2$ and $q=10010_2$ find the values of *a)* $p+q$, *b)* $p-q$, giving your answers in base 2. *c)* If $r=11_2$ show that r is a factor of both p and q. (Work in base 2. Do not convert to base 10.)

8 ABC is a triangle right angled at B. $BC=8$ cm and $\angle ACB=45°$. X is a point in AB such that $AX:XB=2:3$. Calculate *a)* the length AB, *b)* the angle XCB, *c)* the area of triangle ACX.

C8

1 Given that $p=4$, $q=-\frac{1}{2}$, $r=-5$ and $s=\frac{3}{4}$ find the values of

a) $p-r$ *b)* $q+s$ *c)* pq *d)* q/s *e)* $s-qr$ *f)* $\dfrac{2pr}{q}$

2 Using the sets shown in the diagram,

a) list the members of
A $A\cap B$ $A\cap B\cap C$ $A'\cap C$ $(A\cup B\cup C)'$ $A\cup C'$
b) give the value of $n(A)$
c) state whether each of the following is true or false:
i) $6\in A$,   *ii)* $8\in C'$,   *iii)* $\{3,6\}\subset C$,
iv) $\{2,4\}\subset (B\cup C)'$,   *v)* $A'\cap (B\cup C)'=\phi$.

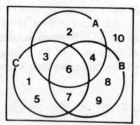

3 The median of 5 integers is 8 and the mode is 6. If the mean is 9 list 4 possible sets of integers.

4 *a)* The area of a triangle is $24\,\text{cm}^2$ and its base is $12\,\text{cm}$. What is its height?
b) ABC is a triangle in which $AB=BC=7\,\text{cm}$, and $AC=12\,\text{cm}$. Find the height of B above AC and the area of the whole triangle.

5 During a period of acute shortage of raw materials, the prices of some manufactured goods were raised by 30%. After the shortages were over, prices were reduced by 20%. If an article was originally priced at £120, what was its price after the shortages were over? Would it have been cheaper if its price had been raised by 10% at the start of the shortages, and left unchanged at the end?

6 Find the radius of a sphere of volume $250\,\text{cm}^3$. Take π as $3\cdot14$.

7 On graph paper, using $2\,\text{cm}$ to one unit on both axes and numbering the x axis from -4 to 3 and the y axis from -2 to 5, plot the three points $(-1,4)$, $(0,3)$ and $(2,4)$. These three points are vertices of a polygon which has $y=1$, $x=-1$ and $x+y=0$ as lines of symmetry. Complete the polygon and give the equations of any other lines of symmetry. What is the order of symmetry of this polygon? Calculate its area. (*Hint* Box it in.)

8 If $P=2x+1$ and $Q=4-3x$, find

a) P if $x=2$ *b)* Q if $x=-2$ *c)* x if $P=6$,   *d)* P if $Q=7$ *e)* Q if $P=4$.
Solve for x: *f)* $P=Q$ *g)* $2P+3Q=0$.

186

C9

1 State whether the following are true or false:

a) $2\cdot2 \times 10^5$ is greater than 2 million.
b) The number of square centimetres in a square metre is 10^4.
c) If the scale of a map is 2 cm to 1 km then an area of $2\,cm^2$ represents an actual area of $1\,km^2$.
d) $\sqrt{0\cdot0026}$ is approximately 0·05.
e) $5\cdot2 \times 10^{-3}$ is greater than $0\cdot25 \times 10^{-2}$.

2 Find the volume of wood left when a cube of side 10 cm has a hole of radius 2 cm bored through its centre. (The hole goes from the centre of one face to the centre of the opposite face.) Take π as 3·14.
Find also the total surface area of the remaining solid.

3 On graph paper, using one large square as the unit on both axes, plot the points (0,0), (2,2) and (3,1). Join up the points to form triangle T. Pre-multiply the position vectors of the vertices of triangle T by

a) Matrix $A \begin{pmatrix} 0 & 1 \\ 1 & 0 \end{pmatrix}$ b) Matrix $B \begin{pmatrix} 0 & 1 \\ -1 & 0 \end{pmatrix}$

c) Matrix A followed by matrix B.

Show the three transformed positions of T on the original graph and describe each of the three transformations geometrically.

4 a) What is the probability of i) throwing a six with one die, ii) throwing a number less than 4 with one die, iii) throwing a total of 10 with two dice?
b) The probability of taking a lemon-flavoured sweet from a bag is 1 in 5 and the probability of taking a lime-flavoured sweet is 1 in 3. If the only other flavour is orange, what is the probability of taking an orange-flavoured sweet?

5 a) The graph shows the distance moved by a body in a given time. Explain the lines OA, AB and BC.

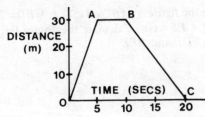

b) The graph shows the speed of a body moving in a straight line for 30 secs.
i) If the acceleration for the first 10 secs is $2\cdot5\,m/s^2$ find x.
ii) Find also the distance covered in the last 20 secs.

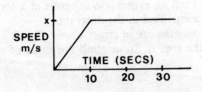

6 O is the centre of the circle.
PQ is parallel to OR.
If angle QOR is $70°$ find the angles QPR and QRP.

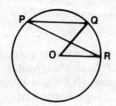

7 a) Using flow charts find the inverse of each of the following functions in the rational domain:

i) $f:x \rightarrow 4-3x$ ii) $f:x \rightarrow \dfrac{6}{x+1}$

b) If $f:x \rightarrow (x-3)(x+1)$ find the value of $f(0)$ and the values of a for which $f(a)=0$.

8 If $X=3a-5$ and $Y=a^2+1$, find   a) X if $a=2$,   b) Y if $a=-2$,   c) a if $X=1$,   d) X if $Y=10$ (two answers).
Solve the equation $X+Y=0$. There are 2, and only 2, sets of values for X and Y.

C10

1 Express 82^2-45^2 as the product of two numbers.

2 If $f:x \rightarrow 3x-2$ and $g:x \rightarrow 4-x$ in the rational domain, find

a) $f(3)$ b) $g(-2)$ c) $f^{-1}(a)$ d) $f^{-1}(4)$ e) $gf(b)$ f) $gf^{-1}(c)$
g) the value of y so that $f(y)=g(y)$.

3 For which of the following would 0·06 be a good approximation?

a) $\dfrac{4·17}{69}$ b) $\sqrt{0·0358}$ c) $11·8 \times 0·051$ d) $0·245^2$

4 Find the slopes of the lines joining these pairs of points:

a) (2,1), (3,2)   b) (0,3), (2,2)   c) (-1, -2), (1, -1)   d) (4,1), (1,4)
e) (1, -2), (0,0)

5 In the figure $\angle ABE=115°$, $\angle CBD=20°$ and $\angle CEB=60°$. Calculate the angles CED, CDB, BDE and BFE.

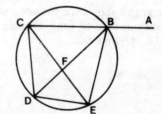

6 Find an expression in terms of x for the area enclosed in the diagram.
All lengths are in cm.
If the area is $20\,\text{cm}^2$, find the value of x.

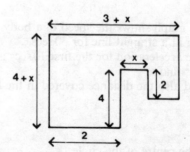

7 $ABCD$ is a rectangular plane in which $AB=6\,\text{cm}$ and $AD=5\,\text{cm}$. The plane is inclined at 40° to the horizontal. If X and Y are points vertically below C and D respectively and $ABXY$ is a horizontal plane, calculate a) CX, b) the inclination of the diagonal AC to the horizontal.

8 Bread is delivered from a bakery in a van which cannot hold more than 30 trays. A tray carries either 14 tin loaves or 8 fancy loaves and the van does not start out on its round with less than 280 loaves.

a) If there are x trays of tin loaves and y of fancy loaves, write down two inequalities relating x and y.

b) If in addition the baker knows that on a certain delivery there must be at least twice as many tin loaves as fancy loaves, write a third inequality.

c) Show on a graph the region where all possible values of x and y are found. Find from your graph the maximum and minimum numbers of loaves that can be carried on that delivery if all the trays are full.

C11

1 In the diagram AB is 6 cm, BC is 5 cm and angle B is 90°.

a) Find the length of AC.

b) If D divides AB in the ratio $AD:DB = 1:2$ write down the length of BD. Calculate $\angle BCD$, $\angle BCA$ and $\angle ADC$.

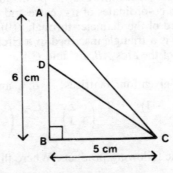

2 a) If $\mathscr{E} = \{1,2,3,4,5,6\}$, $A = \{1,3,4\}$ and $B = \{4,5,6\}$, list the sets A', $A \cap B$, $(A \cup B)'$, $A' \cap B$. What is $n(A \cup B)$?

b) If $\mathscr{E} = \{Dogs\}$, $A = \{White \ dogs\}$, $B = \{Dogs \ over \ 5 \ years \ old\}$, write in words   i) $A \cap B = \phi$,   ii) $A' \neq \phi$,   iii) $B \subset A$.

3 On graph paper number both the x and y axes from 0 to 6. Show by shading the region defined by $y + 3x \geqslant 6$, $x + y \leqslant 5$ and $y \geqslant x$.
Find the least possible value of x which satisfies all three conditions. Find also the maximum value of $2x + y$ and the minimum value of $x + y$.

4 Factorise   a) $3xy + 6y$,   b) $3x^2 - 12x$,   c) $x^2 - 5x - 14$.

5 a) If 2 coins are tossed together, what is the probability of getting either 2 heads or 2 tails?

b) If 4 coins are tossed together, what is the probability of getting 2 heads and 2 tails?

c) If 2 dice are thrown together, what is the probability of scoring a double?

6 AB and CD are diameters of a circle whose centre is O. If angle CAB is 35°, find the following angles:

a) ACD b) AOC c) CDB d) DCB e) COB.

7 Solve the following equations:

a) $x^2 - 4x + 3 = 0$ b) $x^2 - x - 20 = 0$.

1 Mrs Anderson has a bag of 24 ten-pieces for her electric light meter. In an idle moment she noticed that 8 of these are dated 1975 and 11 are dated 1974. What is the probability that the first one she takes out of the bag at random will be dated 1975? If she puts 2 into the meter, and both of these are dated 1975, what is the probability that the next one taken out is dated neither 1974 nor 1975?

2 Draw the x and y axes for values from -3 to $+4$ and show the region for which $x+y \leqslant 4$, $y \leqslant 2x+1$ and $y \geqslant \frac{1}{2}x-2$.
Write down the co-ordinates of the points of intersection of the 3 boundary lines and also the maximum value of $x+2y$ for points which satisfy the given conditions.

3 If $N(x)$ denotes the integer nearest to $\frac{1}{3}x$, e.g. $N(6\cdot2)$ is 2 and $N(-8)$ is -3, find

    a) $N(12)$,   *b)* $N(5)$,   *c)* $N(1)$,   *d)* $N(-10)$,   *e)* $N(-3\cdot5)$.

Given that $N(y)=7$, complete the following for the possible values of y:____ $\leqslant y <$ ____

4 *a)* A circle passes through the points $(0,0)$, $(5,0)$ and $(0,12)$.
Find the co-ordinates of its centre and the co-ordinates of the point which is at the other end of the diameter through $(0,0)$.
    b) ABC is a triangle inscribed in a circle. Find the angles of the triangle if the lengths of the arcs AB, BC and CA are $3b$, $2b$ and $4b$ respectively.

5 You are given four matrices: A, B, C and D.

$$A=(1 \quad 0 \quad -1) \qquad B=\begin{pmatrix} 2 \\ -3 \end{pmatrix} \qquad C=\begin{pmatrix} 2 & 0 & 1 \\ -1 & 2 & 3 \\ 0 & -2 & 1 \end{pmatrix} \qquad D=\begin{pmatrix} 2 & 4 \\ -1 & 3 \\ 0 & 5 \end{pmatrix}$$

Calculate the following products. Where this is not possible say so.

AB AC AD CB DB CD DC

6 AB is a diameter of circle ABC whose centre is O. The tangents at B and C meet at T and BA produced meets the tangent at C at S. If $\angle BOC=114°$, calculate angles

    a) CAO, *b)* CBT, *c)* BCT, *d)* ACS, *e)* CSA.

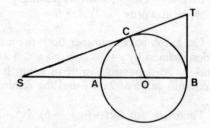

If the radius of the circle is $5\,cm$, calculate the length of the chord CB and the length of the tangent TB.

7 Make a table to show the set of numbers $\{1 \quad 5 \quad 7 \quad 11\}$ under the operation of multiplication (mod 12).
Use your multiplication table to solve the following equations (mod 12)

    a) $5x=7$,   *b)* $\dfrac{5}{y}=11$.

8 A rectangle has a length of $(2x+3)\,cm$ and a width of $(x+1)\,cm$.
Write in terms of x an expression for its perimeter and another expression for its area. If the area is $15\,cm^2$, find the value of x and also the perimeter.

1 In the diagram. *AB* is a diameter of a circle of radius 5 cm. The radius *OD* is perpendicular to *AB*. *ED* is a tangent to the circle at *D* and $\angle DEB = 32°$.

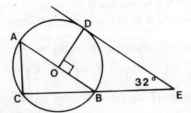

a) What type of quadrilateral is *OBED*?
b) Find $\angle CAB$ and the length of *AC* (to 2 s.f.).
c) The tangent to the circle at *B* cuts *DE* at *F*. Calculate *EF* and the area of *OBED* (both to 2 s.f.).

2 Draw the graph of $y = 2x^2 - 6x - 7$ for values of *x* from -2 to $+5$.
Use your graph to solve the equations a) $2x^2 - 6x - 7 = 0$, b) $2x^2 - 6x = 0$.

3 Find the volume and surface area of the solid which consists of a cylinder of length 8 cm capped by a hemisphere which fits exactly to one end of the cylinder and makes the total length 13 cm.
Take π as 3·142 and give your answers correct to 3 significant figures.

4 On graph paper mark the *x* axis with values from 0 to 8 and the *y* axis from -4 to $+4$. Plot the points *A* (2,1), *B* (3,2), *C* (2,3) and *D* (1,2). Join them up to make a quadrilateral *Q*.

a) Transform *Q* by the matrix $\begin{pmatrix} 1 & 2 \\ 0 & 1 \end{pmatrix}$ and plot the image *R* on the same axes. Describe the transformation.

b) Find the areas of both *Q* and *R*.

c) Transform *Q* by the matrix $\begin{pmatrix} 1 & 0 \\ 0 & -1 \end{pmatrix}$ and plot the image *Q'*.

Using the same matrix transform *R* on to *R'* and mark this also on the graph.
Show the positions of *A*, *B*, *C* and *D* on each of the images.
d) Write down the matrix which maps *R'* on to *Q'*.

5 Given that $\dfrac{90 \times 19}{36} = 47.5$ find the value of: a) $\dfrac{0.9 \times 1.9}{3.6}$ b) $\dfrac{9 \times 0.19}{360}$

6 If $f : x \to \sqrt{x}$ find the value of: a) $f(25)$, b) *a*, where $f(a) = 100$, c) $\dfrac{f(500)}{f(5)}$

7 A local newspaper has columns of 'small ads'. Each advertisement is the same width but can be 2 cm or 3 cm deep. A whole column cannot be longer than 30 cm.

a) If the editor uses *x* '2 cm ads' and *y* '3 cm ads' write down an inequality relating *x* and *y*.
b) It is also decided that in any one column there must be more than three '3 cm ads' and at least as many shorter ones as longer ones. Write down two more inequalities in *x* and *y*.
c) Draw a graph and shade the region in which *x* and *y* must lie. Dot in the grid points in this region. Find from your graph two ways in which a column can be filled.
d) If a '2 cm ad' costs £1.80 and a '3 cm ad' costs £2.40, calculate the maximum amount the paper can collect from one full column.

1   *a*) Find the values of $5x^2 - 2x - 11$ when *i*) $x = -1\frac{1}{2}$, *ii*) $x = 1\frac{1}{2}$.
   b) Draw the graph of $y = 5x^2 - 2x - 11$ for values of x from -2 to $+2$.
   Using your graph, solve the equations *i*) $5x^2 - 2x - 11 = 0$, *ii*) $5x^2 - 2x - 3 = 0$.
   For what range of values of x is $5x^2 - 2x - 11$ less than -4?

2   Identify the transformations represented by the matrices:
$$A = \begin{pmatrix} 1 & 0 \\ 0 & -1 \end{pmatrix} \qquad B = \begin{pmatrix} -1 & 0 \\ 0 & -1 \end{pmatrix} \qquad C = \begin{pmatrix} -1 & 0 \\ 0 & 1 \end{pmatrix}$$
Calculate the matrix product AC and describe the transformation it represents.

3   Find the volume of the solid shown in the figure. It consists of two identical cones, one on each end of a cylinder. Take π as 3·14 and give your answer in cubic metres correct to 2 s.f.

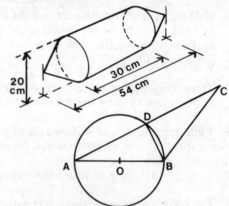

4   In the diagram, AB is a diameter of the circle and triangle ABC is isosceles with $\angle ACB = 34°$ and $AD = 10$ cm. Find *a*) $\angle CAB$, *b*) $\angle ABC$, *c*) $\angle ADB$, *d*) $\angle DBC$, *e*) the area of triangle ACB.

5   If $a*b$ means 'square a and add $2b$', find the values of $4*1$ and $3*(2*1)$. Find x if $5*x = 9$.   Find the two values of y for which $y*y = 15$.

6   A ship sails from a port T for 50 km due East to reach an island S. It then sails 60 km on a bearing $030°$ to island D. Calculate the direct distance of D from T and the course on which the ship should sail to return from D direct to T.

7   The following figures give the daily rainfall, to the nearest mm, measured over a period of 3 weeks:

18, 0, 10, 15, 12, 0, 0, 5, 20, 25, 12, 6, 32, 21, 8, 0, 0, 12, 5, 32, 12

Find *a*) the mode and the median of this set, *b*) the average daily rainfall, *c*) the probability that the Bank Holiday Monday during this period was dry, *d*) the percentage of the total number of days that had more than 15 mm of rain.

8   At a fête the organiser of a competition decides to offer all winners the choice of two prizes, either a packet of crisps or a pen. He does not know how many winners there will be, but decides to have at least 30 prizes.

   a) If he buys x packets of crisps and y pens as prizes, write down an inequality in x and y.
   b) The organiser also decides to have fewer pens than packets of crisps. Write down another inequality.
   c) If crisps cost 5p a packet and pens cost 10p, and he does not want to spend more than £3 on prizes altogether, write down another inequality in x and y.
   d) Draw a graph and show by shading the region in which x and y must lie.

If the pens and packets of crisps can only be bought in multiples of 10, dot in the possible solutions. Which of these is the cheapest? Which provides the largest number of prizes?